Applications of Regression for Categorical Outcomes Using R

This book covers the main models within the GLM (i.e., logistic, Poisson, negative binomial, ordinal, and multinomial). For each model, estimations, interpretations, model fit, diagnostics, and how to convey results graphically are provided. There is a focus on graphic displays of results as these are a core strength of using R for statistical analysis. Many in the social sciences are transitioning away from using Stata, SPSS, and SAS, to using R, and this book uses statistical models which are relevant to the social sciences. *Applications of Regression for Categorical Outcomes Using R* will be useful for graduate students in the social sciences who are looking to expand their statistical knowledge, and for quantitative social scientists due to its ability to act as a practitioners' guide.

Key Features

- Applied – in the sense that we will provide code that others can easily adapt.
- Flexible – R is basically just a fancy calculator. Our programs will enable users to derive quantities that they can use in their work.
- Timely – many in the social sciences are currently transitioning to R or are learning it now. Our book will be a useful resource.
- Versatile – we will write functions into an R package that can be applied to all of the regression models we will cover in the book.
- Aesthetically pleasing – one advantage of R relative to other software packages is that graphs are fully customizable. We will leverage this feature to yield high-end graphical displays of results.
- Affordability – R is free. R packages are free. There is no need to purchase site licenses or updates.

David Melamed is Professor of Sociology and Translational Data Analytics at The Ohio State University. His research interests include the emergence of stratification, cooperation and segregation in dynamical systems, and statistics and methodology. Since 2019, he has been co-editor of *Sociological Methodology*.

Long Doan is Associate Professor of Sociology at the University of Maryland, College Park. His research examines how various social psychological processes like identity, intergroup competition, and bias help to explain the emergence and persistence of social stratification. He focuses on inequalities based on sexuality, gender, and race.

Applications of Regression for Categorical Outcomes Using R

David Melamed and Long Doan

CRC Press
Taylor & Francis Group
Boca Raton London New York

CRC Press is an imprint of the
Taylor & Francis Group, an **informa** business

A CHAPMAN & HALL BOOK

Front cover image: © David Melamed

First edition published 2024
by CRC Press
2385 NW Executive Center Dr, Suite 320, Boca Raton, FL 33431

and by CRC Press
4 Park Square, Milton Park, Abingdon, Oxon, OX14 4RN

CRC Press is an imprint of Taylor & Francis Group, LLC

© 2024 Taylor & Francis Group, LLC

ISBN: 978-0-367-89463-4 (hbk)
ISBN: 978-1-032-50951-8 (pbk)
ISBN: 978-1-003-02984-7 (ebk)

DOI: 10.1201/9781003029847

Typeset in Palatino
by codeMantra

Contents

List of Figures

List of Tables

Acknowledgments

David Melamed thanks Kraig Beyerlein, Ronald L. Breiger, C. André Christie-Mizell, and Scott R. Eliason for their instruction on these topics. Long Doan thanks J. Scott Long for teaching him these models. He also thanks Trent Mize for helping think through some of these issues and valuable citation suggestions. Collaborations with Natasha Quadlin have greatly pushed our thinking on these topics. She also deserves thanks for being an enabler and encouraging the mountain motif in the introduction and on the cover. We also thank Jeremy Freese for his contributions in this area and for encouraging the development of this work. Finally, we thank our editor, David Grubbs, for his encouragement, persistence, and support.

1

Introduction

Many of the processes and outcomes social scientists study are not continuous. From whether a candidate gets hired (binary), to how strongly someone supports a policy position (ordinal), to whether someone goes to restaurant A, B, or C (nominal), to how many cats someone adopts from a local shelter (count), there are countless examples of categorical and count outcomes that social scientists seek to understand and model in their everyday work. We know from theory and practice that the processes underlying these outcomes do not operate linearly. Yet, linear regression models still dominate the empirical landscape. Perhaps, some social scientists find comfort in the simplicity of linear models. Once they are fit, model interpretations come directly off the regression output. The models discussed in this book are more complicated in their estimation and interpretation. But linear regression models applied to categorical and count outcomes can lead to biased and nonsensical predictions. Accordingly, we hope that this book provides a practical guide to understanding and analyzing regression models for categorical and count outcomes.

This book provides a practical initial guide for several classes of regression models that may be intimidating at first. Whereas ordinary least squares (OLS) regression is the de facto standard for linear regressions, approaching nonlinear models like the ones discussed in this book can feel like approaching the foot of a mountain and seeing a landscape featuring many peaks. Our aim is to conceptually explain how to choose which peak to tackle given an empirical problem, distinguish among truly different options, and equivocate when choices are more preferences than substantive decisions. This is not a formal treatment of the statistical models underlying these methods. There are better treatments out there (e.g., Agresti 2003; Arel-Bundock 2022; Eliason 1993; King, Tomz, and Wittenberg 2000; Lüdecke 2018; Mize 2019; Rubin 2004; Cameron and Trivedi 2013; Hilbe 2011; Hosmer and Lemeshow 2000; Long 1997; Train 2009), and we do not feel like we can improve upon them. Instead, it is a book about best practices for data analysis using these methods using R.

Continuing our mountain climbing motif,[1] we hope to not only make the climb less intimidating but also to enable an easy ascent. Once you reach the top of the maximum likelihood curve and have regression output, there is much more ground left to traverse. Unlike a linear regression, much of the work that is to be done with categorical and count models is done postestimation. Through the tools and examples provided, we hope to enable the applied researcher to enjoy the view on the way down and focus on the interpretation and substantive work rather than the monotonous coding work that can make postestimation a chore.

Motivation

This is not the first treatment of these topics, nor will it be the last. Many in the social sciences are moving toward R and other flexible languages for their empirical work. For all its benefits, which we discuss in greater detail in the next chapter, R is not user-friendly like legacy statistical packages such as SAS, SPSS, or Stata. Given the problem of an already complicated set of models, the complexity of R can exacerbate these issues. For both of us, our introduction to these models and a go-to reference for categorical and count regression models is the Long and Freese (2006) book, *Regression Models for Categorical Dependent Variables Using Stata*. It provided a practical guide and very useful functions for interpreting these models but is only developed for Stata. We hope that this can become the Long and Freese adaptation for R.

   One of the downsides of the flexibility of R and its many packages is having to compile a collection of these packages for data analysis. Unlike legacy statistical packages, all of the functions are not built into the core installation of R. There is no "built-in" function to estimate an ordinal regression model, for example. In writing this book, we collate the various functions required for effective analysis and interpretation of categorical and count outcomes. Like Long and Freese (2006), where gaps exist in the toolkit, we provide potential solutions. One such solution is the `catregs` package developed through the writing of this book, which provides similar tools to those provided in Long and Freese's (2006) `SPost` package for categorical dependent variables.

Audience

We write this book with first- and second-year graduate statistics seminar students in mind. To provide an ideal case, this is someone who has already taken an undergraduate statistics class and perhaps a graduate-level linear regression class. We assume that the reader is familiar with the estimation and interpretation of linear regression models. Note that we do *not* assume that the reader is familiar with the underlying math behind these models. Indeed, we find that many practitioners of statistical methods do not need to know the math as proficient computer users do not need to know how to code software or how to build a computer. Of course, we cannot avoid math when writing a regression book. What we emphasize throughout is the *logic* of the math and what these models are doing *conceptually*. We provide enough information to understand what each model is doing and what to watch out for when interpreting these models. Because of our focus on the logical and conceptual underpinning of these models, we highly suggest pairing this book with a more thorough treatment of the theory and statistics behind the model for readers who really want to understand them.

Likewise, we assume that the reader is familiar with using R for linear regressions. We walk through the syntax and options for each function call throughout the book, but a superficial familiarity with the basics of object-oriented programming will greatly help in understanding these functions and R more generally.

Coverage and Organization

This book covers the most commonly used models for cross-sectional analyses of categorical and count outcomes. For binary outcomes, where the outcome only has two values (e.g., yes or no, died or lived, hired or not hired), we discuss the binary logit and binary probit models. For ordinal outcomes, where the outcome takes on multiple categories that are assumed to be ordered on some underlying dimension (e.g., strongly disagree to strongly agree, not at all to all the time, highest educational attainment), we discuss the ordered logit, ordered probit, and the partial proportional odds model. For nominal outcomes, where the categories are not ordered (e.g., mode of transportation, political affiliation, marital status), we focus on the multinomial logit model. For count outcomes, where the outcome is the number of times something has happened (e.g., number of children, number of cats, number of sexual partners), we begin with the Poisson regression model and introduce various complications through the negative binomial model, the zero-inflated model, the zero-truncated model, and finally the hurdle model. We also cover some commonly used but less neatly categorizable models like the conditional logit model and the rank-ordered logit model. Finally, we end with two special topics – cross-model comparisons and missing data. Both required at least some attention for a complete treatment of these models.

The book can be thought of as being divided into two parts: the preparation stage and the execution stage. In the first part of the book, we walk through the underpinning of the models discussed in this book and the necessary legwork to get to the analysis stage. In the second part of the book, we walk through the models of interest. Chapter 2 begins with an overview of R Studio and a discussion of the packages used throughout the book. Here, we provide not only a list of necessary packages but also an explanation of why certain packages are preferred where choices exist. This is also where we introduce and walk through the core functionality of the `catregs` package. Chapter 3 begins with OLS regression and transitions to the generalized linear model. In this chapter, we provide the conceptual linkages between the model readers are probably most aware of and the models we introduce them to later in the book. Chapter 4 discusses useful tests of associations and approaches to basic descriptive statistics for categorical outcomes. We find that many students want to jump head-first into a complicated model when it is necessary to first understand the distribution of responses and the data.

In the second half of the book, we walk through the models of interest. Chapter 5 discusses regression for binary outcomes, focusing on the binary logit model. Because this model forms the conceptual core of many of the other models discussed in this book, this chapter is a deeper treatment of the underlying logic of the model and its interpretation. Chapter 6 continues with more in-depth "advanced" topics of interpretation, including moderation and squared terms. The methods and strategies discussed in Chapters 5 and 6 carry on throughout the rest of the book. Chapter 7 discusses regression for ordinal outcomes, focusing on the ordered logit model. We end this chapter with a discussion on relaxing the assumptions of the ordered logit and transition to the multinomial logit. Chapter 8 discusses regression for nominal outcomes, focusing on the multinomial logit model. Chapter 9 discusses regression for count outcomes, starting with the Poisson regression model and adding various complications from overdispersion to zero-inflation and zero-truncation. Finally, the last two chapters cover miscellaneous topics that are important but do not neatly fit in elsewhere. Chapter 10 discusses additional outcome types like ranks and covers the rank-ordered logit as well as the conditional logit and Chapter 11 discusses cross-model comparisons and missing data.

Note

1. We should probably note that neither of us are mountain climbers so the metaphor may not be perfect.

2

Introduction to R Studio and Packages

R is a free, comprehensive, and inclusive computing environment. On the one hand, this means that it is flexible, more efficient than many alternatives, and does things that stand-alone statistics packages do not do (e.g., natural language processing or web scraping). On the other hand, it means that there are no computer programmers who develop graphical user interfaces that account for what users want from their statistical packages. That is to say, there is a trade-off between flexibility and user-friendliness, and R certainly prioritizes flexibility over ease of use.

We implement R in the context of RStudio. RStudio is an integrated development environment (IDE) that functions as a wrapper for base R. Although base R includes many statistical procedures, including many used in this book, it is closer to a calculator than it is to standard statistics packages (e.g., SAS or Stata). RStudio includes base R as part of the environment, but it also includes scripts to maintain and execute your code, integrated help files, and the ability to manage larger projects.

We do not offer an overall introduction to R or to RStudio. There are many wonderful and free resources for learning the basics online. A good introduction to R can be found in Crawley (2012) and Wickham and Grolemund (2016), and a good introduction to RStudio can be found in Verzani (2011). Instead, we assume that our readers are familiar with R and RStudio, and have some limited experience estimating statistics in this context. In this chapter, we describe the additional tools that are required to work with regression models for limited dependent variables. In particular, we discuss objects in R, packages, data visualization in R, and the `catregs` package that we developed for this book.

Objects in R

R is an object-oriented programming language. When you open the software, you have an empty environment that you can populate with as many objects as you have computer memory. Everything you want to save or manipulate at another time needs to be defined as an object in the environment. This includes data files, model objects or results, graphs, and so on. Importantly, you can have as many data files in your environment as you want. That is, unlike standard statistical software, which traditionally only allows analysts to open a single dataset, R enables analysts to work with multiple data files simultaneously.

We often have multiple data files at a time loaded in our environment. As such, we briefly discuss data management in this context. Within project folders, we find it useful to create additional subfolders for data files and possibly for visualizations. Once the folders are defined, R has an option to set your working directory. Once set, that is the default location where R looks for files. This can be accomplished with the setwd command, or you can set the path using the drop-down menu (Session -> Set Working Directory -> Choose Directory), which has the nice feature of printing the code to the console.

R's file extension for data files is .RData. You can save one or more objects in such a file using the save function (see: ?save). Aside from saving data, we point out that you can save model results, graphs, or any object using this function. This is particularly useful when models take a long time to converge. Such results can be saved and reused later. Aside from RData files, we also find comma separated values (.csv) to be an effective data format when working in R. There are base R commands to import and export csv files (read.csv and write.csv, respectively), and most other packages can directly import such files as well. R can directly import data from stand-alone statistical software packages but doing so requires downloading a package that is not built into the base R installation. We discuss packages in the next section.

Packages

Base R includes many mathematical and statistical functions. For example, the function to estimate the models we discuss in Chapter 5 is available in R by default (i.e., glm). R also allows users to write their own functions (see: ?function). If we define a function and load it into our environment, we can apply that function to other objects in our workspace. Consider a simple function that exponentiates numbers and adds five. Here is the R code to define such a function in the environment, and then apply it to the numbers zero through five:

```
> f1 <- function(x){
+    x2 <- exp(x) + 5
+    return(x2) }
> f1(0:5)
[1]    6.000000    7.718282   12.389056   25.085537   59.598150  153.413159
```

In the above, we begin by initializing the `function` and the objects that will be manipulated go inside the parentheses. In this case, there is only one variable to be manipulated, but this can be a list of objects to manipulate. The brackets define the function itself. In this case, we apply a simple mathematical transformation to the input, `x`, to define the output, `x2`. Finally, we specified what the function should `return`. Here again, this is a single object, but this can be a list of items to return as well.

One of R's great strengths is the ability to create *packages* out of functions that other analysts can access. Anyone can write new functions, put them into an R package, and make that package available to all. Packages can include one or more functions, help files associated with each function, data files, vignette examples of package features, and/or additional ancillary files. Many such packages are hosted on CRAN (Comprehensive R Archive Network) and can be installed via code (see: `?install.packages`) or via drop-down option (`Tools -> Install Package`). Packages on CRAN have been evaluated, ensuring they work across platforms for example. Packages are often available to users before they get archived on CRAN. It is common for developers to post their packages on Github. Such packages can be installed using `devtools::install _ github` function (Wickham et al. 2021).

One of the workhorse packages that we rely upon throughout this book is the `tidyverse` package (Wickham et al. 2019). `Tidyverse` is a suite of packages developed for data science in R. When you load the `tidyverse` package [e.g., `require(tidyverse)` following `install.packages("tidyverse")`], it loads eight different packages, including `dplyr` and `ggplot2`. If you are unfamiliar with tidyverse, we recommend spending some time working with the verbs of data science in R, namely, `mutate`, `group _ by`, `filter`, and `summarize`. A good resource is Chapter 5 of *R for Data Science* (Wickham and Grolemund 2016). In isolation, these functions are very useful. But when "piped" together (Wickham and Grolemund 2016), they create an intuitive workflow for manipulating, grouping, and summarizing data.

Visualization is one of R's great strengths. Across data science platforms, `ggplot2` is one of the more intuitive and flexible data visualization suites available. Here again we assume some familiarity with this package, but we also provide all our code online. Good introductions to `ggplot2` can be found in Wickham, Chang, and Wickham (2016) and Healy (2018). The object-oriented nature of R means that everything from a model object to raw data can be manipulated, transformed, and/or reshaped, and saved as a different object. And most objects can be fed to `ggplot2` to generate publication-quality graphics that are 100% developed and reproducible with code. All model-based figures in this book were developed using `ggplot2`. We also note that `ggpubr` (Kassambara 2020) makes combining graphs into paneled figures very straightforward.

Another package worth mentioning in this context is the `foreign` package (Team et al. 2022). It has functions to read and write data files from SAS,

SPSS, and Stata. We do not use this package in this book, but it is something that we commonly use in our work. The `stargazer` package (Hlavac 2015) is also worth mentioning. It creates formatted tables out of one or more model objects in either HTML or LaTex format. The formatted tables can then be placed directly in word processing software (e.g., TexShop or Word via html formatting).

In terms of estimating the models in this book, we rely on several packages. The `MASS` package (Ripley et al. 2013) includes functions for estimating ordinal logistic regression (Chapter 7) and negative binomial regression (Chapter 9). It also includes the `mvrnorm` function that takes random draws from a multivariate normal distribution. We use this function for simulation-based inference of predicted values. Chapter 8 uses both the `nnet` package (Ripley, Venables, and Ripley 2016) and the `mlogit` package (Croissant 2012) for multinomial logistic regression. In Chapter 9, we use the `pscl` package (Jackman et al. 2015) to estimate zero-inflated count models and hurdle models, and the `countreg` package (Zeileis, Kleiber, and Jackman 2008) to estimate zero-truncated count models. Chapter 10 implements the `Epi` package (Carstensen et al. 2015) for conditional or fixed effects logistic regression and the `mlogit` package (Croissant 2012) for rank-ordered or exploded logistic regression. Finally, in Chapter 11, we describe the `mice` package for multiple imputation using chained equations for missing data (Van Buuren and Groothuis-Oudshoorn 2011). We use several other packages throughout and try to give credit when we do.

Once we have estimated a regression model, we need to interpret it. Particularly with the models discussed in this book, doing so usually entails visualizing the implications of the model. For example, we plot one of our independent variables on the x-axis and show how the predicted values from the model changes over its interval on the y-axis, after setting covariates to their means or other meaningful values. There are some excellent post-estimation options in R, and we have developed several new functions that we describe in the next section. But first we describe existing packages that are used throughout this book.

The `emmeans` package (Lenth 2021) generates estimated marginal means or least-square means from model objects. We give it the values to use for the independent variables and it computes the predicted values (e.g., predicted probabilities in the case of logistic regression) along with their delta method standard errors (Agresti 2003). The output from `emmeans` is then tweaked slightly (e.g., adding labels) and plotted using `ggplot2`. A nice extension to `emmeans` is the `ggeffects` package (Lüdecke 2018). It automates generating fitted values and plot creation, but there are fewer options for customization. For this reason, we think analysts are better off working directly with `emmeans` and `ggplot2`, but note that `ggeffects` is an efficient approach for more exploratory analyses.

Aside from predicted values, we are often interested in the marginal effects of our independent variables – either on average, for every respondent in the data, or with covariates to set to some meaningful values. There are two R packages that compute marginal effects: the `margins` package (Leeper 2017) and the `marginaleffects` package (Arel-Bundock 2022). We rely on the `marginaleffects` package for computing average marginal effects throughout this book. We also use the `deltamethod` function within the `marginaleffects` package to compute standard errors on differences in predicted values.

The catregs Package

The `catregs`[1] package contains 11 functions that we implement throughout this book. The core of the package is six functions for working with the fitted values of model objects. The functions define a pipeline for both generating sets of predicted values and estimates of uncertainty, and for testing or probing how those predicted values vary. The `margins.des` function creates design matrices for generating predicted values, the output of which is fed to `margins.dat`. `margins.dat` is a wrapper for `emmeans`. Each row of the design matrix is automatically fed to an `emmeans` statement to generate fitted values and estimates of uncertainty. It automatically detects the type of model based on its class and generates predicted values on the response scale, along with their delta method standard error.

For testing first and second differences in fitted values, we developed `first.diff.fitted` and `second.diff.fitted`.[2] These are both wrapper functions for the `marginaleffects::deltamethod` function. Our functions simply streamline the implementation of probing multiple comparisons. More specifically, `first.diff.fitted` computes the difference in fitted values and the standard error from pairs of rows in the design matrix generated by `margins.des`. Multiple comparisons may be generated by listing more than one pair of rows to compare (see, e.g., Chapter 6). Similarly, the `second.diff.fitted` function facilitates computing second differences in fitted values. It takes a model object, a design matrix, as generated by `margins.des`, and an ordered set of four rows to compare, and it returns the second difference and its standard error. Collectively `margins.des`, `first.diff.fitted`, and `second.diff.fitted` make probing interaction effects in the generalized linear model in R more straightforward.

Generally, there are three approaches to estimating the uncertainty in margins or fitted values from models (Mize 2019). The functions described above rely on the delta method (Agresti 2003) and bootstrapping. An alternative approach is to simulate the uncertainty in marginal effects (King, Tomz, and Wittenberg 2000). The general `compare.margins` function relies on simulation-based inference. Given two marginal effects and their standard errors, the `compare.margins` function simulates draws from the normal distributions implied by the estimates and computes a simulated *p*-value that refers to the proportion of times the two simulated distributions overlapped. This function can compare any two marginal effects, as estimated by `margins.dat`, `marginaleffects`, `first.diff.fitted`, `second.diff.fitted`, and so on. It is a flexible way to probe effects. We caution readers to ensure that they *should* be comparing the marginal effects they put in this function. For example, cross-model comparisons of marginal effects require adjustments (see Chapter 11).

Models for alternative-specific outcomes, including conditional logistic regression and rank-ordered logistic regression (Chapter 10), are not supported by emmeans. We developed the `margins.dat.clogit` function to generate estimates of uncertainty associated with predicted probabilities from these models. By default, the function uses simulation to generate a sampling distribution and returns both the inner 95% (by default) of the simulated distribution and the standard deviation of the distribution. To do so, it simulates random draws from a multivariate normal distribution (Ripley et al. 2013) that is defined by the estimated coefficients and their variance/covariance matrix. Those estimates are used to generate predicted probabilities, and this process is repeated many times (1,000 by default). Simulation is an efficient means to inference, but we caution readers to ensure that they simulate a large enough number of fitted values. In practice, we find it useful to rerun our results with different seeds, comparing between solutions. It may be necessary to increase the number of simulations for the solutions to converge, particularly for large or complicated model specifications.[3] Simulation-based inference is the default for conditional logistic regression and rank-ordered logistic regression. For conditional logistic regression, bootstrapping is also supported. The supporting documentation in the package describes how bootstrapping is implemented and how to alter the default settings, such as the number of samples to use.

Aside from the six functions described above, `catregs` includes five other functions that we use throughout the book. `lr.test` is a basic implementation of the likelihood ratio (LR) test for comparing nested models (Eliason 1993). The base R function `anova` will compute an LR test but only reports *p*-values for models that were estimated with maximum likelihood. As discussed in Chapter 3, there are two primary ways to estimate the regression coefficients in regression models for limited dependent variables – maximum likelihood and iteratively reweighted least squares (IRLS). Unfortunately, the `glm` function uses IRLS, meaning that LR tests for logit or Poisson models do not include *p*-values. Our `lr.test` function takes two models – any two models for which there is a `logLik()` solution, including those estimated with the `glm` function – and reports the chi-squared test statistic, the degrees of freedom, and the associated *p*-value. The order of the models in terms of "full" and "reduced" does not matter.

When working with regression models for limited dependent variables, we are often interested in functions of the estimated regression coefficients.

As discussed in Chapter 5, we do not model limited dependent variables directly; instead, we model some mathematical transformation of them. We often then invert that transformation in post-estimation to understand what the results mean. For example, when modeling a binary variable using logistic regression (Chapters 5 and 6), it is often useful to exponentiate the regression coefficients. The `list.coef` function is an alternative or supplement to the `summary` function for regression models for limited dependent variables. In addition to the estimated regression coefficients and their standard errors, `list.coef` returns exponentiated coefficients, percent change, and the lower and upper limits of confidence intervals.

Case-level diagnostics for regressions of limited dependent variables have some support in R. We developed the `diagn` function to automate much of what is available. The function detects the model type based on its `class`, and then automatically computes the relevant diagnostic statistics, such as standardized residuals. The supplemental code can be adjusted with minimal effort to automate visualizing your model diagnostics (e.g., changing the name of your "data" file).

In the context of regression models for count outcomes (Chapter 9), Stata's `countfit` function (Long and Freese 2006) summarizes a lot of information. In that context, there are several possible models that might be preferred, and there are empirical benchmarks for deciding between the models. The `countfit` function estimates four different models of the count process, prints their results, along with overall summaries to the console, and generates a graph of model residuals. As it is a very useful function, we emulated it in our `count.fit` function. We describe this function in more detail in Chapter 9.

The final function that is included with `catregs` is `rubins.rule`. As the name implies, the function implements Rubin's rule for combining standard errors (Rubin 2004). The method was developed for use in the context of multiple imputation, and we use it in that context as well (Chapter 11). But Rubin's rule is also useful for combining standard errors from multiple estimates of marginal means, as these are assumed to be normally distributed as well. For example, we use Rubin's rule to combine standard errors when computing average marginal effects at means in Chapter 6.

Conclusion

While R is perhaps less user-friendly than some other packages (e.g., Stata), it is flexible, free, can be updated at any time, and has functionality that exceeds or rivals the best modern data science platforms (e.g., Python), particularly in data management (`dplyr`) and data visualization (`ggplot2`). At the time of this writing, we were able to estimate all the models we wished to cover in this book with available packages. In Chapter 11, we discuss one instance where further development is needed (i.e., seemingly unrelated estimation). Available R packages do not offer a lot of support for alternative-specific models (Chapter 10). If your data structure requires modifications to a standard conditional logistic regression set-up, you will likely be better off transitioning to a different software platform (e.g., SAS and Stata both have more support in this space). Hopefully others can fill these gaps with new packages in the coming years.

Packages facilitate open science. As new methods are developed, developers can allow others to implement their methods by publishing packages. And packages evolve, with updated versions as functions are added or adjusted. The next iteration of `catregs` will support multilevel or mixed effects regression models for limited dependent variables. Generally, we recommend that readers check for updates periodically and install the latest versions of packages. Of course, it is also worthwhile to periodically search for new packages that might make your analyses even more efficient.

Notes

1. As of this writing, `catregs` is not on CRAN. It is available on Github: `devtools::install _ github("dmmelamed/catregs")`.
2. Supported models for `first.diff.fitted` and `second.diff.fitted` include all models covered in Chapters 3–9. Partial proportional odds models, zero-inflated models, zero-truncated models, and hurdle models are only supported with nonparametric inference via bootstrapping. Other models are supported with parametric inference via delta method standard errors and nonparametric inference via bootstrapping.
3. "Converge" here means some threshold that is smaller than you will be using in your public-facing presentation of results. If confidence intervals will use three decimals, for example, ensure your simulated standard errors are changing on a smaller scale than that from run to run.

3

Overview of OLS Regression and Introduction to the Generalized Linear Model

In this chapter, we review the ordinary least squares (OLS) regression model. For more in-depth, theoretical treatments, we recommend Gunst and Mason (2018) or Neter and colleagues (1996). We review the model itself, its interpretation, and how to implement the model in R. We then discuss how to alter the model for noncontinuous dependent variables using the link function, resulting in the generalized linear model. Finally, we discuss parameter estimation in the context of the generalized linear model. The material corresponding to parameter estimation is slightly more technical but is provided for those who want hands-on experience with these models. Readers more interested in the applied aspects of regression in R may wish to skim or skip the sections on parameter estimation.

The OLS Regression Model

In this chapter and throughout this book, we use data from the European Social Survey (ESS). Specifically, we use data from respondents in the United Kingdom from the 2020 wave of the ESS. Figure 3.1 presents the univariate and bivariate distribution for two variables.

The first is generalized trust (A). Specifically, participants were asked to rate whether most people can be trusted, or you can't be too careful, with 10 indicating that people can be trusted. The second variable is education, reflecting years of formal schooling completed (B). Panel C of Figure 3.1 shows a dot plot or scatterplot of the two variables. In the graph, we have "jittered" the points. Jittering means adding a small amount of random noise to points in a graph so that points are not stacked on top of one another. In this case, both variables are measured in discrete integer units, so there are multiple respondents at the intersection of each variable. This does not show up with a traditional graph, that is, without jittering. But with jittering, we can see that there is a small positive association between the variables. As a prelude to the regression results, we have included the best fitting linear line between the two variables in the figure. It highlights the positive relationship

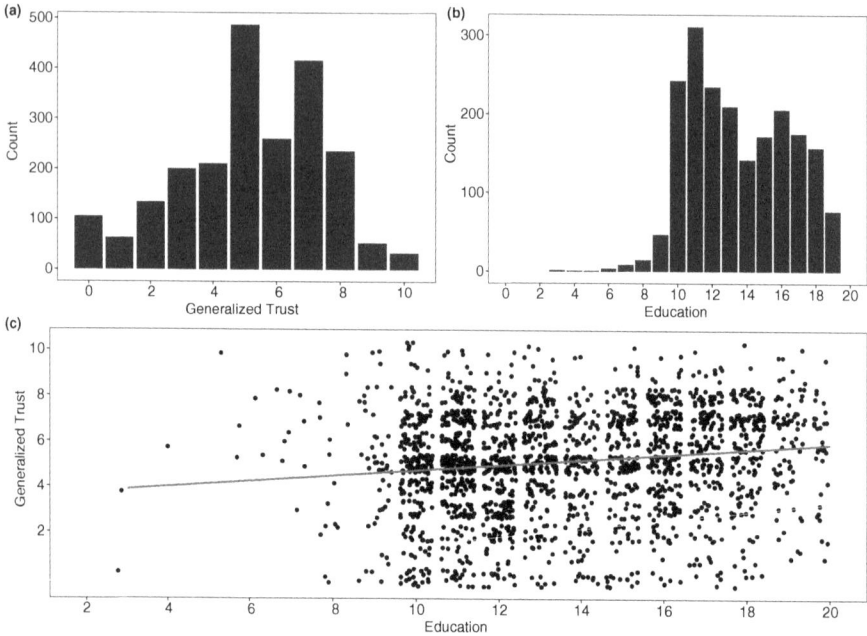

FIGURE 3.1
Univariate distributions of generalized trust (a), education (b), and the bivariate distribution of the two (c).

in the scatterplot. OLS regression formalizes the return to generalized trust for a unit increase in education.

Table 3.1 summarizes two OLS regression models predicting generalized trust from the ESS. Model 1 is a *bivariate* regression model, predicting generalized trust as a function of only education. The estimated slope of the line in Figure 3.1 is .102, the slope of the regression equation. It means that, on average, for each unit increase in education, there is a corresponding .102 unit increase in generalized trust. As one's education increases, so does one's generalized trust ($\beta = .102$, $p < .001$). The R-squared value indicates that education explains 2.6% of the variation in generalized trust. The final reported statistic for Model 1 is the global F-statistic, which tests whether the model as a whole is more predictive than using the mean of the response. This statistic is useful for model comparisons, as described below.

Of course, one of the reasons for the proliferation of linear statistical models in social sciences is their ability to *control for* other factors. There may be some confounding third factor that affects the relationship between education and generalized trust, such as age. For example, age may make people more or less trusting, or alter one's assessment of trust. Regression allows analysts to statistically control for the covariance between variables, isolating the independent effect of each variable on the outcome of interest. Model 2 in Table 3.1 adds controls for religiousness, age, being racially minoritized, and sex. Introducing these controls enhances the effect of education slightly,

TABLE 3.1

Summary of Two OLS Regression Models

	Model 1	Model 2
Education	0.102***	0.115***
	(0.013)	(0.014)
Religious		0.057***
		(0.017)
Age		0.009**
		(0.003)
Minority (=1)		−0.447*
		(0.189)
Female (=1)		−0.098
		(0.098)
Constant	3.741***	2.959***
	(0.196)	(0.274)
R²	0.026	0.042
F	$F_{(1,2150)} = 57.40$	$F_{(5,2146)} = 18.63$

Note: Source is the European Social Survey. $^{*}p<.05$, $^{**}p<.01$, and $^{***}p<.001$.

indicating that at least one of the variables suppresses the effect of education (MacKinnon, Krull, and Lockwood 2000). We also find that more religious ($\beta=.057$, $p<.001$), older ($\beta=.009$, $p<.01$), and nonminoritized ($\beta=.447$, $p<.05$) respondents report higher life satisfaction. Adding these four control variables increases the explained variance by .016 over Model 1.

Regression Estimates

Equation 3.1 presents the OLS regression model. The response is y_i, independent variables are represented by x_k, and a stochastic component is represented as the model residual, ε_i. The subscript i represents the observation number, from 1 to N. The regression coefficients, $\beta_0 \ldots \beta_k$, indicate the weight of the corresponding factor on the response. β_0 is the model intercept and is the expected value of the response when all predictors are set to zero. The same formula may be written more concisely in matrix notation. Equation 3.2 is identical to Eq. 3.1, but it is in matrix notation. Per convention, lower-case letters denote vectors ($n\times1$) and upper-case letters denote matrices ($n\times k$). In Eq. 3.2, **X** includes a vector of 1's to represent the model constant, but the constant may be suppressed if the response has been mean-centered. The coefficients or weights from the OLS model may be obtained using linear algebra (Neter et al. 1996). Equation 3.3 presents the linear solution to the regression coefficients. In Eq. 3.3, $\mathbf{X}^{\mathsf{T}}$ is the notation for the transpose of matrix **X**, and $\mathbf{X}^{-1}$ is the notation for inverting matrix **X**. In this context, we identify the OLS estimator (Eq. 3.3) in order to describe how it is generalized for regression models with limited dependent variables.

$$y_i = \beta_0 + \beta_1 x_{i1} + \beta_2 x_{i2} + \ldots + \beta_k x_{ik} + \varepsilon_i \tag{3.1}$$

$$\mathbf{y} = \mathbf{Xb} + \varepsilon \tag{3.2}$$

$$\mathbf{b} = \left(\mathbf{X}^\mathsf{T}\mathbf{X}\right)^{-1}\mathbf{X}^\mathsf{T}\mathbf{y} \tag{3.3}$$

The regression coefficient is the weight associated with a 1 unit change in the dependent variable for a 1 unit change in the independent variable. The standard error is an estimate of the standard deviation of that effect or weight. The standard error is defined as the square root of the diagonal of the variance/covariance matrix of the coefficients. The variance/covariance matrix is defined as:

$$VCOV(\mathbf{b}) = MSE\left(\mathbf{X}^\mathsf{T}\mathbf{X}\right)^{-1}$$

MSE refers to the mean squared error and is defined as:

$$MSE = \frac{\Sigma\left(y_i - \hat{y}_i\right)^2}{N - (p+1)}$$

$\hat{y}_i$ refers to the fitted or predicted values from the regression equation and p refers to the number of estimated parameters (i.e., the length of $\mathbf{b}$). Conceptually $\left(\mathbf{X}^\mathsf{T}\mathbf{X}\right)^{-1}$ quantifies the covariance between the variables and the MSE quantifies the average error of prediction. Their product therefore tells us about the variance of the estimates. The statistical significance of those estimates is determined by computing the probability that the slope could be zero in the population given the data at hand. Specifically, we assume that the ratio of the coefficient to its standard error is t-distributed, with degrees of freedom equal to the sample size minus the number of parameters plus 1 (denominator of the MSE). Under this assumption, we interpret the p-value as the probability that the observed effect is zero in the population.

The R^2 statistic tells us how much variation in the dependent variable is explained by the independent variables. Like analysis of variance, this is accomplished by partitioning the total outcome variance into parts explained and unexplained by the model. The total variance or total sum of squares (TSS), the unexplained or residual sum of squares (RSS), and explained or model sums of squares (MSS) are defined as:

$$TSS = \Sigma\left(y_i - \bar{y}\right)^2$$

$$MSS = \Sigma\left(\hat{y}_i - \bar{y}\right)^2$$

$$RSS = \Sigma\left(y_i - \hat{y}_i\right)^2$$

TSS quantifies the squared errors of prediction when only conditioning on the mean of the outcome. MSS quantifies the squared errors reduced when conditioning on the entire model and RSS quantifies the squared errors remaining after conditioning on the model. Given these, the explained variance is defined as:

$$R^2 = MSS / TSS = 1 - RSS / TSS$$

Adding variables to an OLS regression model increases the explained variance. Continuing to add variables to an OLS regression model continues to increase the explained variance, even if those variables are not adding to the predictive power of the model. As such, we often report an adjusted R^2 that accounts for the number of terms in the model. The adjusted R^2 can decrease as uninformative predictors are added to a model. The adjusted R^2 is defined as:

$$R^2_{Adj} = 1 - \left[\frac{\left(1 - R^2\right)(N - 1)}{N - p} \right]$$

The final statistic reported in Table 3.1 is the model F-statistic. The F-statistic is used to compare two *nested* models, meaning that the *full* model includes all model terms in a *reduced* model, plus some additional terms. The model F-test that is typically reported with OLS regression output compares the estimated model to a null model with no predictors. Generally, the F-statistic may be defined as:

$$F = \frac{\left(RSS_R - RSS_F\right) \Big/ k}{MSE_F}$$

With DF$=k$, N$-p_f$, where the R and F subscripts denote statistics from the reduced and full models, respectively, and k denotes the difference in the number of parameters between the two models. The statistic has DF equal to k and n $-p_f$, where p_f refers to the number of estimated parameters in the full model. The statistic reported in Table 3.1 compares each of the models reported to a null model, using only the mean of the response for prediction. The test is useful more generally for comparing estimated models. For example, we can test whether adding the entire block of control variables to Model 2 adds significantly to the predictive power of Model 1. The analogous test of nested models for regression models with categorical or limited dependent variables is used quite often throughout the subsequent pages of this book.

Once the model is defined, we use it to generate predictions. The fitted values from the model are defined above as $\hat{y}_i$. Up to this point, predictions are those implied by the independent variable matrix – the predicted value for the individual respondent in the data. But model predictions are often used to illustrate the implications of the model. We create some ideal typical data, for example, with everyone at the mean on all variables except for one and we systematically vary that one variable. Such predictions show how the response varies over the predictor, net of all the controls in the model (e.g., Figure 3.3). Very often, model predictions are used to illustrate statistical interactions, with covariates set to specific values or their means (see below). To know whether the patterns in such plots are statistically significant, we need estimates of uncertainty for such point estimates. Estimates of uncertainty for fitted values do not come from model estimates themselves but must be computed post-estimation (Mize 2019).

Common approaches for estimating uncertainty around model predictions include the delta method (Agresti 2003), statistical simulation (King et al. 2000), and bootstrapping (Efron and Tibshirani 1994). The delta method is a parametric approach to obtaining estimates of uncertainty (Agresti 2003). The delta method uses a Taylor Series expansion of the linear predictors to approximate the margin in the neighborhood of the estimated coefficients and provided independent variable values. To do so, it differentiates the function via a Taylor Series expansion (i.e., the formula for the predicted value) and pre- and post-multiplies it by the model variance/covariance matrix. In short, the delta method aggregates over the precision of estimates in the model, and the amount of information in the model at the point of the margins. Agresti (2003: 576–581) provides more technical details on the delta method. The simulation approach relies on the assumption that the estimated regression coefficients are multivariate normally distributed. Given this assumption, which we make in estimating the OLS regression model, we can simulate a draw from the multivariate normal distribution using the observed coefficients and their variance/covariance matrix. These simulated values can be used to compute predicted or fitted values. If we repeat this enough times, we generate a sampling distribution of predicted or fitted values. The estimates from the repeated simulation procedure are treated as a sampling distribution for inference: the inner 95% of the distribution defines a simulation-based confidence interval around the model prediction. Bootstrapping is accomplished by repeatedly sampling from the original data some subset of the sample (with replacement), reestimating the model, reestimating the predicted values, and repeating the process many times. Significance is determined in the same fashion as it is for the simulation approach. The `catregs` package relies on the delta method and simulation techniques, but bootstrapping is an option for many models.

Estimating Regression Models in R

R has a base function, `lm`, that estimates linear regression models. This function computes everything we described above, and beyond. As a best practice, we recommend creating a model object in your R workspace. You can manipulate it later, compare between models, create tables with several models for exporting, and so on. For example, the call to estimate Model 1 of Table 3.1 and then see the model output is as follows:

```
m1 <- lm(can.trust.people ~ education ,data=X)
> summary(m1)
Call:
lm(formula = can.trust.people ~ education, data = X)

Residuals:
    Min      1Q  Median      3Q     Max
-6.3809 -1.3655  0.1422  1.6345  5.7514

Coefficients:
            Estimate Std. Error t value Pr(>|t|)
(Intercept)   3.7409     0.1963  19.055  < 2e-16 ***
education     0.1015     0.0134   7.576 5.25e-14 ***
---
Signif. codes:  0 '***' 0.001 '**' 0.01 '*' 0.05 '.' 0.1 ' ' 1
Residual standard error: 2.262 on 2150 degrees of freedom
Multiple R-squared:  0.026,      Adjusted R-squared:  0.02555
F-statistic:  57.4 on 1 and 2150 DF,  p-value: 5.249e-14
```

Everything in our summary of the model in Table 3.1 is included in this query for a summary of the model object. It is also useful to see the names of the other objects inside the model:

```
> names(m1)
 [1] "coefficients"  "residuals"       "effects"   "rank"
"fitted.values" "assign"
 [7] "qr"                  "df.residual"     "xlevels"   "call"
"terms"               "model"
```

You can index into lm model objects to access everything listed above. The coef or vcov functions of base R can also extract the coefficients and variance/covariance matrix for the coefficients, which is needed to derive the standard errors. For example, here are the model coefficients and standard errors for Model 2:

```
> coef(m2)
(Intercept)    education    religious         age     minority       female
 2.95892145  0.11459992  0.05688131  0.00900213 -0.44696936 -0.09761066
> sqrt(diag(vcov(m2)))
(Intercept)    education    religious         age     minority       female
0.274111841 0.013658906 0.016833008 0.002815825 0.189217766 0.098295629
```

Note that the last expression combines three functions. The square root sqrt() is applied only to the diagonal entries diag() of the variance/covariance matrix of the coefficient vcov(m2).

Interpreting Regression Models with Graphs

Coefficient plots are an alternative to tables summarizing regression output. For example, Figure 3.2 is a coefficient plot illustrating the results of Model 2. You can quickly see that all of the variables except sex are predictive of generalized trust, that education seems to have a comparatively strong effect, and minoritized individuals show decreased generalized trust.

The supplemental R code illustrates how to generate the coefficient plot. We use the broom package to turn the model object into something amenable for plotting. We then remove the constant, due to its arbitrary nature. We then used ggplot in tidyverse to plot the results.

As noted above, another approach to graphing regression results is to illustrate model predictions as key variables vary over some interval. For example, Figure 3.3 illustrates how predicted values of generalized trust vary as education varies. It clearly shows that increases in education results in increases in generalized trust. Such plots are ubiquitous in the social sciences, particularly to illustrate the patterns implied by interaction effects. We will see such examples in this chapter. The process used to generate Figure 3.3 is very common workflow. Given the model at hand, we created an ideal typical data set that sets all covariates to their means (or other useful values) while systematically varying education. Here is the data set we generated:

```
> round(design,2)
   education religious    age minority female
1          8        3.6 53.24     0.08   0.55
2          9        3.6 53.24     0.08   0.55
3         10        3.6 53.24     0.08   0.55
4         11        3.6 53.24     0.08   0.55
5         12        3.6 53.24     0.08   0.55
6         13        3.6 53.24     0.08   0.55
7         14        3.6 53.24     0.08   0.55
8         15        3.6 53.24     0.08   0.55
9         16        3.6 53.24     0.08   0.55
10        17        3.6 53.24     0.08   0.55
11        18        3.6 53.24     0.08   0.55
12        19        3.6 53.24     0.08   0.55
13        20        3.6 53.24     0.08   0.55
```

After defining our ideal typical data, we then estimated marginal means or fitted values from the regression model. The emmeans package (Lenth 2021) in R, which is called by the catregs package when computing parametric standard errors, computes delta method standard errors for model predictions. Finally, we plot the model predictions, including the estimate of uncertainty as a 95% confidence interval.

As we have noted, it is common to present plots of fitted values from regression models to illustrate statistical interactions. Table 3.2 includes two regression models, both estimated with statistical interactions in the models. Model 1 shows that the effect of education varies by religiousness. There is

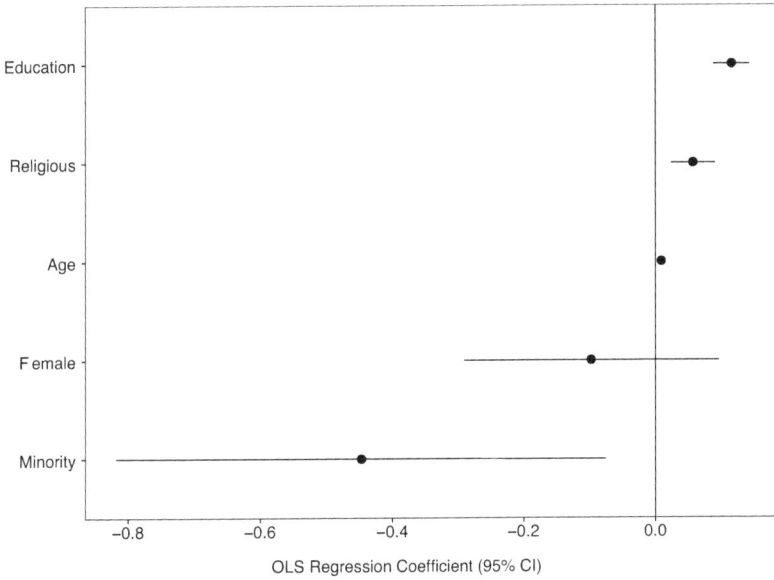

FIGURE 3.2
Coefficient plot summarizing an OLS regression of generalized trust on several predictors.
Coefficients are denoted by points and error bars denote 95% confidence intervals.

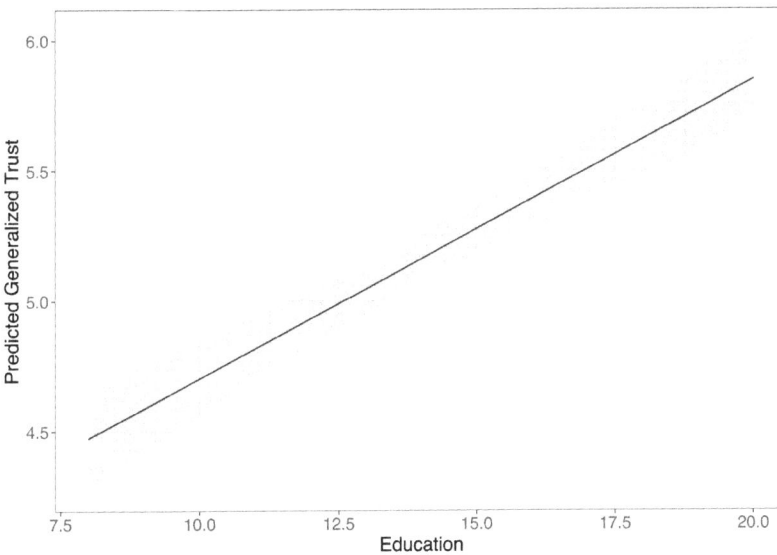

FIGURE 3.3
Estimated marginal generalized trust as a function of education. Estimates drawn from Model
2 of Table 3.1. Covariates set to their means. 95% confidence intervals (delta method) reported.

a positive main effect of education, a positive main effect of religiousness,
and a negative interaction effect. Collectively these three terms imply that
as religiousness increases, the return to education decreases. But it is hard to

"see" this relationship by just thinking about the interrelationship between the three estimated coefficients. Figure 3.4 presents marginal means or fitted values from the regression model. We have again set all covariates to their means while generating ideal typical data that systematically varies education and religiousness. Below is the ideal typical data, with education varying from 8 to 20 and religiousness varying from. 54 (the mean minus one standard deviation) to 6.66 (the mean plus one standard deviation):

```
> round(design2,2)
   education religious    age minority female
1          8      0.54 53.24      0.08   0.55
2          9      0.54 53.24      0.08   0.55
3         10      0.54 53.24      0.08   0.55
4         11      0.54 53.24      0.08   0.55
5         12      0.54 53.24      0.08   0.55
6         13      0.54 53.24      0.08   0.55
7         14      0.54 53.24      0.08   0.55
8         15      0.54 53.24      0.08   0.55
9         16      0.54 53.24      0.08   0.55
10        17      0.54 53.24      0.08   0.55
11        18      0.54 53.24      0.08   0.55
12        19      0.54 53.24      0.08   0.55
13        20      0.54 53.24      0.08   0.55
14         8      3.60 53.24      0.08   0.55
15         9      3.60 53.24      0.08   0.55
16        10      3.60 53.24      0.08   0.55
17        11      3.60 53.24      0.08   0.55
18        12      3.60 53.24      0.08   0.55
19        13      3.60 53.24      0.08   0.55
20        14      3.60 53.24      0.08   0.55
21        15      3.60 53.24      0.08   0.55
22        16      3.60 53.24      0.08   0.55
23        17      3.60 53.24      0.08   0.55
24        18      3.60 53.24      0.08   0.55
25        19      3.60 53.24      0.08   0.55
26        20      3.60 53.24      0.08   0.55
27         8      6.66 53.24      0.08   0.55
28         9      6.66 53.24      0.08   0.55
29        10      6.66 53.24      0.08   0.55
30        11      6.66 53.24      0.08   0.55
31        12      6.66 53.24      0.08   0.55
32        13      6.66 53.24      0.08   0.55
33        14      6.66 53.24      0.08   0.55
34        15      6.66 53.24      0.08   0.55
35        16      6.66 53.24      0.08   0.55
36        17      6.66 53.24      0.08   0.55
37        18      6.66 53.24      0.08   0.55
38        19      6.66 53.24      0.08   0.55
39        20      6.66 53.24      0.08   0.55
```

As shown above, values for educations are varied within the three levels of religiousness (mean–1sd, the mean, and mean+1sd). We then computed predicted values for each row of the ideal data, along with estimates of uncertainty using the `margins.dat` function in the `catregs` package. After coding a custom x-axis for the plot, we then used `ggplot` to create Figure 3.4. The plot shows that education has a stronger effect on generalized trust for those who are more religious. That is, religion can act as a buffer on trust for those who are less educated. This interpretation is not readily apparent from examining the regression coefficients alone.

We point out that tests of second differences in predicted values can recover the interaction effect in the context of OLS regression. We revisit this point when we discuss interaction effects in the generalized linear model with noncontinuous outcomes. Specifically, the second differences between adjacent education categories and religiousness are equal to –.011 (the estimated interaction effect). For example, when religiousness is 3, and education is 12, a 1 unit increase in education is associated with a .126 unit increase in generalized trust. When religiousness is 4, a 1 unit increase in education is associated with a. 115 unit increase in generalized trust. The difference between these first differences, that is, the second difference, is estimated to be –.011, which is the same as the interaction effect in Model 1 of Table 3.2. Furthermore, if we compute the delta method standard error for this second difference, we recover the interaction effect and its standard error. In the case of OLS regression, the second difference of adjacent terms in interaction

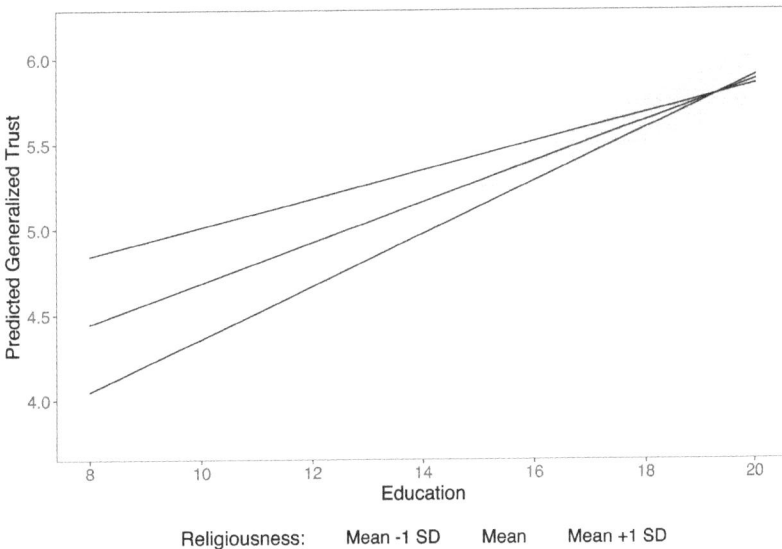

FIGURE 3.4
Estimated marginal generalized trust as a function of education and religiousness. Estimates drawn from Model 1 of Table 3.2. Covariates set to their means. 95% confidence intervals (delta method) reported.

TABLE 3.2

Summary of Two OLS Regression Models
Including Statistical Interaction Effects

	Model 1	Model 2
Education (E)	0.161***	0.124***
	(0.022)	(0.014)
Religious (R)	0.222***	0.056***
	(0.063)	(0.017)
Age	0.009**	0.009**
	(0.003)	(0.003)
Minority (=1)	−0.437*	0.849
	(0.189)	(0.697)
Female (=1)	−0.101	−0.095
	(0.098)	(0.098)
E × R	−0.011**	
	(0.004)	
E × M		−0.084
		(0.043)
Constant	2.326***	2.823***
	(0.361)	(0.283)
R^2	0.045	0.043
F	$F_{(6,2145)}=16.78$	$F_{(6,2145)}=16.17$

Note: Source is the European Social Survey, UK sample.
$^*p<.05$, $^{**}p<.01$, and $^{***}p<.001$.

effects and its delta method standard error reproduces the estimated interaction effect and its standard error. We point this out here because tests of second differences have become the norm for assessing interaction effects in the generalized linear model (Mize 2019). We pick this point back up throughout subsequent chapters.

Diagnostics

Of course, regression models rely on several assumptions, including that the observations are independent from one another, that the error term follows a normal distribution, and that the error term has constant variance over levels of independent variables. It is also important to assess whether individual cases exert undue influence on the model. Note, however, that we do not provide an in-depth treatment of diagnostics for OLS models. Belsley, Kuh, and Welsch (2005) provide a good, in-depth treatment of these issues. Instead, below we review those diagnostics that are relevant to the generalized linear model.

A good place to start when assessing an OLS regression model is to look at a scatterplot of model residuals by fitted values from the model. This can identify high-level problems with the model, such as poor predictive performance at high or low levels of the response, which results in less variability and characteristic signs of heteroscedasticity. Heteroscedasticity refers to uneven variability in the model residuals, typically with a conic shape. Figure 3.5a shows such a plot for Model 1 in Table 3.2. There appears to be no concern with heteroscedasticity for this model. Across levels of the response there appears to be relatively uniform variability. At higher levels it is harder to make out but jittering the points in this example would alter the interpretation. To generate this plot, we use the `augment` function in the `broom` package (Robinson, Hayes, and Couch 2022) to append model estimates (residuals, fitted values, etc.) to the observed data. We then directly plot that object using `ggplot2`.

It is also common to inspect model residuals and their variability at observed levels of key independent variables. If model residuals are uniformly small, on average, across levels of key independent variables, it is a sign of efficient estimation (Snijders and Bosker 2011). Figure 3.5b shows such a figure, illustrating that the observed residuals vary depending on the level of education. This is a sign of a poorly specified model – the residuals should be close to zero regardless of the level of education. To generate this plot, we aggregated the residuals from the `augment` function. Specifically, we used the `group _ by` function in `tidyverse` to group the data by levels of education and used `summarize` to compute the mean and standard deviation. The results were then directly plotted using `ggplot` to generate Figure 3.5b.

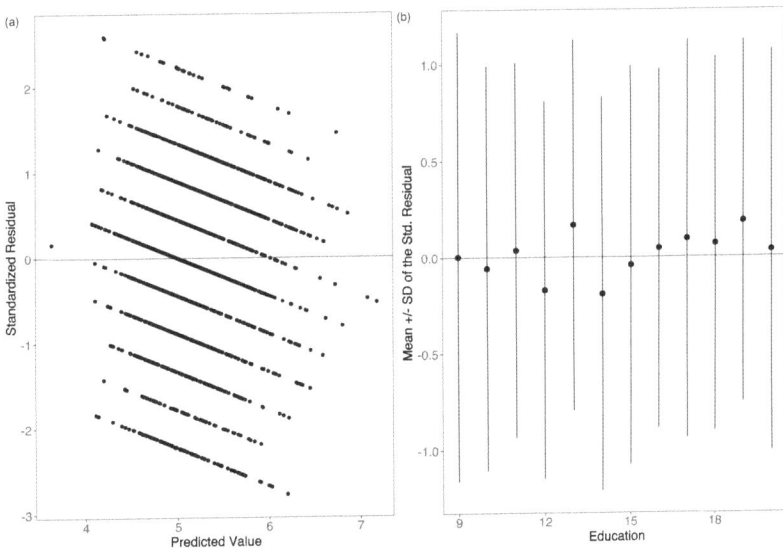

FIGURE 3.5
Scatterplot of predicted values by standardized residuals for Model 2 of Table 3.1 (a) and a dot-and-whisker plot of standardized residuals at observed levels of education (b).

The Generalized Linear Model (GLM)

The GLM and Link Functions

We quickly run into problems applying OLS regression to categorical and limited responses. Consider a binary response. First, nonsensical predictions, such as predicted probabilities above 1.0 and below 0.0, are possible. Second, OLS regression assumes constant error variance across levels of independent variables (see above), but OLS models with a binary response have heteroscedasticity (nonconstant error variance; Long 1997: 38–39). With a binary response, the residual is mathematically tied to the linear prediction. Smaller residuals are observed for smaller or larger probabilities and larger residuals are observed for probabilities near .5. For these reasons, we prefer to model a transformation of the outcome, rather than the outcome itself. Specifically, in the GLM, we assume a functional form of our dependent variable that is continuous, just like the right-hand side of the regression equation. More generally, the generalized linear model is expressed in Eq. 3.4, and more compactly in matrix notation in Eq. 3.5. We point out two key differences between the GLM presented here and the OLS regression equation in Eq. 3.1. First, in Eq. 3.4, we use $\eta(y_i)$ to denote that we are modeling a *function* of the response, not the response itself. Second, due to modeling a function of the response, we make differential assumptions about the model residual depending upon how we transform our response. We say more about this in subsequent chapters as we describe specific models.

$$\eta(y_i) = \beta_0 + \beta_1 x_{i1} + \beta_2 x_{i2} + \ldots + \beta_k x_{ik} \qquad (3.4)$$

$$\eta(y) = \mathbf{Xb} \qquad (3.5)$$

The function on the response is termed the *link function*. The link function transforms the response so that it has a continuous interval, rather than a categorical or discrete interval. Consider a binary outcome as a function of a continuous prediction equation. Figure 3.6a illustrates directly modeling the binary response, assuming it is continuous. This is termed a *Linear Probability Model*. As shown in the graph, and noted above, predictions above and below the reasonable range for probabilities are generated at low and high values of the predictor. Alternatively, instead of modeling the binary response, social scientists often model the logit or log-odds of the response. Figure 3.6b illustrates how the predicted values vary over the continuous interval when the logit of the response is used as the link function. More generally, using the logit link function ensures that model predictions are greater than zero and less than one. The generalized linear model that assumes a binomial distribution with a logit link function is more generally known as *logistic regression*. Chapters 5 and 6 discuss logistic regression in depth.

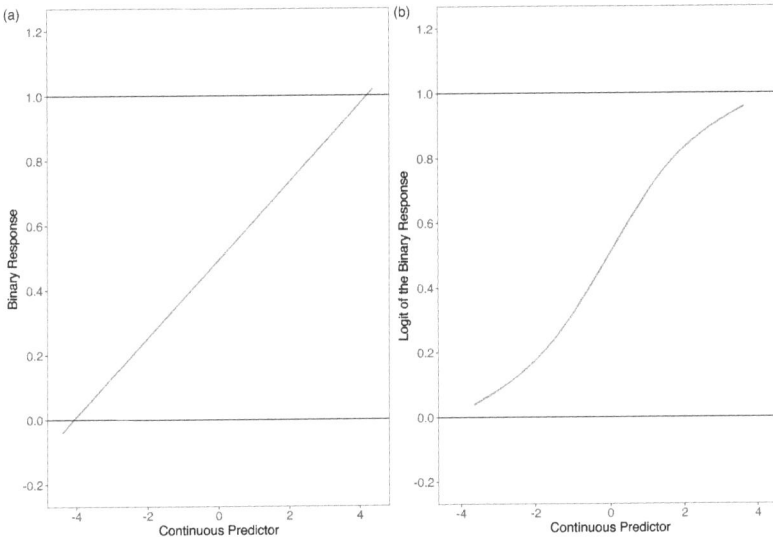

FIGURE 3.6
Illustration of identity and logit link functions applied to a binary response over a continuous interval.

The introduction of the link function into the regression equation breaks the linear relationship between the dependent variable and the predictor variables or predictor matrix. As such, we rely on iterative solutions to solve for the regression coefficients in the GLM. Below, we describe two such procedures – iteratively reweighted least squares and maximum likelihood estimation. But first, we describe how to generate predicted values from the GLM.

The regression equation for the GLM (Eqs. 3.4 & 3.5) says that the logit of the response is a linear function of the predictors. To derive predicted values from the model, we apply the inverse of the link function to both sides of the equation. This transforms the linear prediction (i.e., **Xb**) into the constrained response metric of the link function. In this example, with a binary response and a logit link function, we apply the *logistic* function to the linear prediction to generate the predicted probabilities. We describe this process for each model type we consider in subsequent chapters.

Iteratively Reweighted Least Squares (IRLS)

R's glm function relies on IRLS to solve for model coefficients. In general terms, IRLS (*i*) estimates an OLS regression model to use as start values, using the estimator in Eq. 3.3, (*ii*) uses those values to generate predicted values and to compute a working response or pseudovalues (e.g., Hilbe 2011: 52; Hosmer and Lemeshow 2000: 129), and (*iii*) then re-estimates the coefficients using weighted regression, with some function of the fitted values serving as the weights and the working response used as the outcome. Steps *ii* and *iii* are repeated until the coefficients stop changing (within some *tolerance* or threshold of zero). The working response and the function of the fitted values changes depending on the assumed distribution of the outcome. We do not cover the details or formulas here, but the supporting R code illustrates IRLS for logistic regression, and we refer readers to Hosmer and Lemeshow (2000) and Hilbe (2011) for more details on IRLS in the context of logistic regression and regression for count models, respectively. Their notation should be familiar, as it has influenced our own.

Maximum Likelihood (ML) Estimation

The ML estimates are the parameter estimates that, if true in the population, would be the most likely to generate the observed data. For a given model using ML, there is a corresponding function that computes the likelihood or probability of the observed data given the model parameters. The likelihood function varies with the link function used (Agresti 2003: 136). For simple models, the solution to this function is the point at which the derivative of the likelihood function equals zero (Agresti 2003: 10). More complicated formulations rely on iterative procedures. For example, the Newton–Raphson algorithm is implemented in several popular R packages. It is an iterative procedure where the ratio of the first-order derivatives of the log-likelihood function (termed the Gradient) to the second-order derivatives of the log-likelihood function (termed the Hessian Matrix) is subtracted from the coefficient estimates at each iteration. Once the coefficients stop changing from iteration to iteration, it has identified the ML solution. We refer readers to Eliason (1993) for more on the derivation of likelihood functions and identifying their maximums. The supplemental R code illustrates ML using base R's `optimize` function to solve the solution to the log-likelihood function for logistic regression coefficients.

Model Comparison in the Generalized Linear Model

In the context of OLS regression, the *F*-test for nested models is a useful tool for assessing whether multi-category variables add to the overall predictive power of the model or for testing whether adding blocks of terms (e.g., a set of controls) adds to the model. Similarly, the likelihood ratio (LR) test of nested models enables us to assess whether the difference between two nested models is statistically significant. Both those models must have a solution to a log-likelihood function. More than one parameter may be constrained at a time. Let L_F denote the solution of the log-likelihood function for the full model, and let L_C denote the solution to the log-likelihood function for the constrained model, then the LR statistic is defined as follows (Powers and Xie 2008: 67):

$$G^2 = -2(L_F - L_C)$$

The LR test of nested models can be implemented with the `lm.test` function in the `catregs` package. It is also available in base R's `anova` function, but it only fully works on models estimated with ML. The `lm.test` function is more general in that it makes a call for the solution of the log-likelihoods of the models and directly compares them. As such, `lm.test` can be applied to all the models in this book. For example, below is the `lm.test` output after comparing the OLS models in Table 3.1. Note that the results are asymptotically identical to the *F*-test. While we did not detail the estimation of linear regression via maximum likelihood, suffice it to say that an ML solution exists for model parameters of linear models as well.

```
> lr.test(m1,m2)
    LL.Full LL.Reduced G2.LR.Statistic DF p.value
1 -4809.265  -4791.892        34.74632  4       0
```

When models are nested in one another, we rely on the LR test of nested models (G^2) for model selection, and we recommend that you do as well. However, not all models are nested in one another. More complicated model specifications, particularly those with many variables, may not be strictly nested in one another. When models are not nested, the LR test is inappropriate. Alternatively, one can rely on *information criteria* for model selection. In particular, both the Bayesian and Akaike's Information Criteria are popular metrics for model selection when models are not nested (Akaike 1974; Schwarz 1978). In both cases, smaller values indicate better comparative model fit. The formula for both are provided below. Raftery (1995: 139) notes that a difference in Information Criteria of 10 is roughly equivalent to a significant difference.

$$\text{BIC} = -2 \times L + \log(N) \times npar$$

$$\text{AIC} = -2 \times L + 2 \times npar$$

Where L is the log likelihood for the model and npar is the number of parameters in the model. The primary difference between the two measures is that AIC penalizes the complexity of the model linearly while BIC penalizes it relative to the log of the number of observations. Neither is inherently better and neither measure makes sense when looking at one model. To use either measure, we would need to compare the AIC and BIC across models, with lower values indicating better fit (higher maximized likelihoods relative to the number of parameters used).

4

Describing Categorical Variables and Some Useful Tests of Association

In this chapter, we describe categorical data and the relationships between categorical variables. We begin with univariate statistics and generalize from there. We also provide some inferential tests of association between categorical variables. First, we describe the Chi-squared test of independence for cross-classified data. Then, we describe the more general log-linear model for cross-classified data.

Univariate Distributions

Nominal variables can be meaningfully summarized as the count of respondents in each category or the proportion of the sample in each category. Ordinal variables can be summarized as the count of respondents in each category, the proportion or percent of the sample in each category, and the proportion or percent of respondents in each category plus all the categories above or below that category. That is, because the categories can be ordered, we can also compute the cumulative proportion/percent of the sample above or below each category.

Table 4.1 is a summary of two categorical variables from the United Kingdom's data from the European Social Survey. First, we report the relative sample size of respondents in each response category. Then, we report the percent of respondents in each category. For education we also report the cumulative percent of respondents in each category and every category below that one. Note that this measure is inappropriate for nominal variables, much like it is inappropriate to report the standard deviation of a nominal variable. As shown in Table 4.1, the European Social Survey has responses from 1,206 people who identify as female and 998 people who identify as male. In other words, the sample is 54.7% female. Similarly, Table 4.1 shows that 869 respondents, or 39.6% of the sample, have a high school diploma or less, 524 respondents (23.9%) have some college, 206 respondents (9.4%) have a college degree, and the remaining 590 respondents (27.0%) have a graduate education. The cumulative percent is the last column in Table 4.1. It shows

DOI: 10.1201/9781003029847-4

TABLE 4.1

Description of Sex and Education from the European Social Survey

Sex	n	%	Education	n	%	Cum. %
Female	1,206	54.7	High School or Less	869	39.7	39.7
Male	998	45.3	Some College	524	23.9	63.6
			BA/BS	206	9.4	73.0
			Graduate School	590	27.0	100.0

that 63.6% of respondents have some college or less, and that 73% of respondents have a college degree (BA or BS) or less.

As implemented in the replication script, tabulating a variable in R is accomplished using the `table` command. There are many ways to generate percent of respondents in each response category, but we recommend dividing the tabled variable by the sum of the table (e.g., `table(education)/sum(table(education))`). For ordinal variables, such as education in Table 4.1, the cumulative proportion may also be useful. Once the proportions in each response category are computed, the `cumsum` function can automatically convert them into cumulative proportions. In the supporting material, we illustrate this in base R and in `tidyverse`.

Bivariate Distributions

The European Social Survey asks respondents whether they feel safe walking alone at night. Response categories included "very unsafe," "unsafe," "safe," and "very safe." We binarized this variable into states of unsafe and safe. All tolled 1,654 respondents (75.4%) feel safe walking alone at night. Table 4.2 shows the cross-tabulation of respondent sex by feeling safe walking alone at night. This is also called a contingency table. It shows the count of respondents in the cross-classification of both variables. In the European Social Survey, for example, there are 396 females and 143 males who do not feel safe walking alone at night. Cross-tabulations are generated in R by combining multiple variables into the `table` command (e.g., `table(sex, safe)`).

Table 4.3 presents general notation for joint proportions, conditional proportions, and for the margins. Of course, it is all quite simple since we have a 2×2 table. The joint proportion of the sample that is both female and does not feel safe at night is .181. The joint proportion is the count of respondents in a cross-classified cell, divided by the entire sample. The conditional proportion tells us that 67% of females feel safe at night, compared to 85.6% of males. And the marginal distributions reproduce the univariate distributions of the respective row and column variables.

TABLE 4.2

Cross-Tabulation of Respondent Sex and Whether
They Feel Safe Walking Alone at Night

	Not Safe	Safe
Female	396	803
Male	143	851

Note: Source is the European Social Survey.

TABLE 4.3

Notation and Observed Values for Joint, (Conditional), and Marginal Proportions
from the Cross-Classification of Sex by Feeling Safe Walking Alone at Night.

	Column		Marginal
Row	1 – Not Safe	2 – Safe	
1 – Female	$\pi_{1,1}=.181$	$\pi_{1,2}=.366$	$\pi_{1,+}=.547$
	$\left(\pi_{1\mid1}\right)=.330$	$\left(\pi_{2\mid1}\right)=.670$	$\left(\Sigma\,\pi_{i\mid1}\right)=1.00$
2 – Male	$\pi_{2,1}=.065$	$\pi_{2,2}=.388$	$\pi_{2,+}=.453$
	$\left(\pi_{1\mid2}\right)=.144$	$\left(\pi_{2\mid2}\right)=.856$	$\left(\Sigma\,\pi_{i\mid2}\right)=1.00$
Marginal	$\pi_{+,1}=.246$	$\pi_{+,2}=.754$	1.00

Note: Conditional proportions are denoted by parentheticals. This Table is modeled after
Agresti's (2003: 39) Table 2.3.

The *difference in proportions* shows that males are 18.6% more likely than
females to report feeling safe walking alone at night (i.e., $\pi_{2\mid2}-\pi_{2\mid1}=.186$).
This difference in proportions simply tells us the group difference in the
proportion of 1s. One way to understand the magnitude of this difference is
in relation to the incidence of 1s. In this regard, the *relative risk* is the ratio of
proportions conditional on some categorical variable. In this case, the relative
risk for females is .78 (i.e., $\pi_{2\mid1}\,/\,\pi_{2\mid2}=.67/.856$), indicating that females are at a
lower overall "risk" to feel safe walking alone at night compared to males.
Technically, women are .78 times as likely to report feeling safe. Consider if
feeling safe was lower overall, say 20% for males and 1.4% for females. There
is the same absolute difference, only the relative difference has changed.
Now the relative risk for females to feel safe would be .07, or women would
be .07 times as likely as men to report feeling safe. More generally, as the
difference in proportions increases and as the overall proportion decreases,
the relative risk increases. It is also worth noting that relative risks that are
less than 1 are often inverted so that the interpretation is more intuitive. For
example, "men are 1.28 times more likely to report feeling safe at night (i.e.,
1/.78 or .856/.67)" is more straightforward than "women are .78 times more
likely report feeling safe at night (i.e., .67/.856)."

Next, we discuss the *odds*. Given the proportion of successes or observed 1's, here denoted as π, odds are defined as: $\pi/(1-\pi)$. For example, women are 2.03 times more likely to report feeling safe walking alone at night than unsafe (i.e., $.67/(1-.67)$), or men are 5.95 times more likely to report feeling safe at night than unsafe (i.e., $.856/(1-.856)$). The ratio of proportions provides useful information, namely, the relative risk, the ratio of odds, or *odds ratio*, is quite useful, and shows up in many statistical contexts (see also, Chapters 5–8). The odds ratio can be defined as: $\dfrac{\pi_1/(1-\pi_1)}{\pi_2/(1-\pi_2)}$. Applied to our case, the odds ratio is $\dfrac{\pi_1/(1-\pi_1)}{\pi_2/(1-\pi_2)} = \dfrac{.67/(1-.67)}{.856/(1-.856)} = .34$. This means that the odds that a female feels safe walking alone at night are $.34$ times the odds that a male reports feeling safe. As with relative risk, when odds are <1, we often invert them and change the reference category to ease interpretation. For example, men are 2.93 times more likely than women to report feeling safe walking alone at night (i.e., $1/.34 = 2.93$). In this example, the entire 2×2 Table is needed to compute a single odds ratio. The supporting R script reproduces these quantities (e.g., difference in proportions, odds, and odds ratios, using familiar commands).

Before generalizing these notions to larger and somewhat more realistic tables, we turn to the Pearson (1900) Chi-Squared Test of Independence for contingency tables. The intuition behind the test is clear: if the two (or more) variables of interest are independent from one another, we would expect to see cell counts in contingency tables that reflect only the marginal distributions, that is, a joint product of the margins. The null hypothesis of every Pearson Chi-squared test is that the two (or more) variables contributing to the cross-classification table are independent. The alternative is simply that they are associated. The tools described above, along with inspection of the Chi-squared residuals, can tell where within the bivariate distribution the association occurs.

Let N denote the sample size in a contingency table and let $N_{i,j}$ denote the count of respondents in the *i*th, *j*th cell within that contingency table. For example, $N_{1,2}$ for the cross-classification of sex and feeling safe at night in the ESS is 803 (Table 4.2). Given this, the χ^2 test statistic equals

$$\chi^2 = \sum_i \sum_j \frac{\left(N_{i,j} - N\pi_{i,+}\pi_{+,j}\right)^2}{N\pi_{i,+}\pi_{+,j}} \tag{4.1}$$

The degrees of freedom[1] (DF; Fisher 1922) for the test are equal to the number of rows in the table minus 1 times the number of columns in the table minus 1, commonly denoted $(R-1) \times (C-1)$. In this case, there is 1 degree of freedom. Thus, the χ^2 test statistic would need to exceed 3.84 to be considered significant at conventional thresholds. With 1 degree of freedom, a value of

3.84 leaves 5% of the area under the Chi-squared distribution to the right (e.g., pchisq (3.84,1,lower.tail=FALSE)).

The Pearson χ^2 test statistic is implemented in Base R (e.g.,?chisq.test). By default, it does not use Pearson's original formula. This can be estimated by adding the option "correct=FALSE." By default, R implements a continuity correction that adds a small constant to each residual, which prevents the overestimation of significance in small samples (Yates 1934). Specifically, the default formula for the χ^2 test statistic in the chisq.test command is as follows:

$$\chi^2 = \sum_i \sum_j \frac{\left(\left|N_{i,j} - N\pi_{i,+}\pi_{+,j}\right| - .5\right)^2}{N\pi_{i,+}\pi_{+,j}} \tag{4.2}$$

The χ^2 for Table 4.2 is 101.88 for the original formula and 100.87 for the continuity corrected correct formula. Here is the corresponding R output for both tests:

```
> chisq.test(table(X$gender,X$safe),correct=FALSE)

        Pearson's Chi-squared test

data:   table(X$gender, X$safe)
X-squared = 101.88, df = 1, p-value < 2.2e-16

> chisq.test(table(X$gender,X$safe))

        Pearson's Chi-squared test with Yates' continuity
correction

data:   table(X$gender, X$safe)
X-squared = 100.87, df = 1, p-value < 2.2e-16
```

Using the original or uncorrected formula, we find that the probability of observing the contingency table we did is <.001 if sex and feeling safe at night were really independent. Alternatively, using the corrected formula, the probability of observing the contingency table we did assuming sex and feeling safe at night are independent is also <.001. We have a large sample size so it is unsurprising that the two methods return a similar answer. Either way, we reject the hypothesis that sex and feeling safe at night are independent. Of course, we may want to control for other factors that might be associated with both gender and feeling safe at night, and we will be able to do so in the context of logistic regression (Chapters 5 and 6). In this case, we have a decent sense of the association in the data: males are more likely to report feeling safe at night than females.

Table 4.4 presents a second contingency table from the ESS. While this table also only includes only two variables, it is vastly more complicated than

TABLE 4.4

Cross-Tabulation of Life Satisfaction and Generalized Trust

Satisfied with Life	Can Other People Be Trusted?		
	Few People	Some People	Most People
Not satisfied	140	19	35
A little satisfied	139	30	86
Somewhat satisfied	123	66	135
Very satisfied	142	82	239
Completely satisfied	169	63	243

Note: Source is the European Social Survey, UK Sample.

Table 4.2. In this case, we have two cross-classified ordinal variables. The first is life satisfaction, with five response categories. The second is a measure of generalized trust: how many people can be trusted – "few," "some," or "most." There are five response categories for the row variable and three for the column variable, meaning that there are ten DF for this table. There are also ten unique odds ratios that one could interpret from this contingency table. More generally, there are *DF* unique odds ratios implied by any contingency table.

The components of the χ^2 test statistic can indicate proportional contributions to the breakdown of statistical independence. By components, we refer to the cell-level contributions to the overall χ^2 test statistic. Specifically, we recommend plotting them, in a heatmap fashion, as in Figure 4.1. Of course, the code to replicate the Figure is provided. In each cell of the cross-classified figure, we report the Chi-squared component for the cell. Importantly, although the components are all positive, we coded cells with fewer respondents than expected as negative. Larger negative values (more purple) indicate fewer respondents than expected if life satisfaction and generalized trust were independent, and positive values (more orange) indicate greater respondents than expected. The largest contribution to the breakdown of independence in these data is that there are more people than expected who are not satisfied with life and trust few other people (χ^2 component=43.3). More generally, the heatmap shows that people who are less satisfied with their life are more likely to trust "few" people and are correspondingly less likely to trust "some" or "most" people, and those more satisfied with their life are likely to trust "most" people than "few" people. The ordinal nature of these variables fortunately shows up in the observed trends. The method of interpreting χ^2 components is a nice complement to relative risks, odds, and odds ratios as a means to understanding the patterns of association in a contingency table.

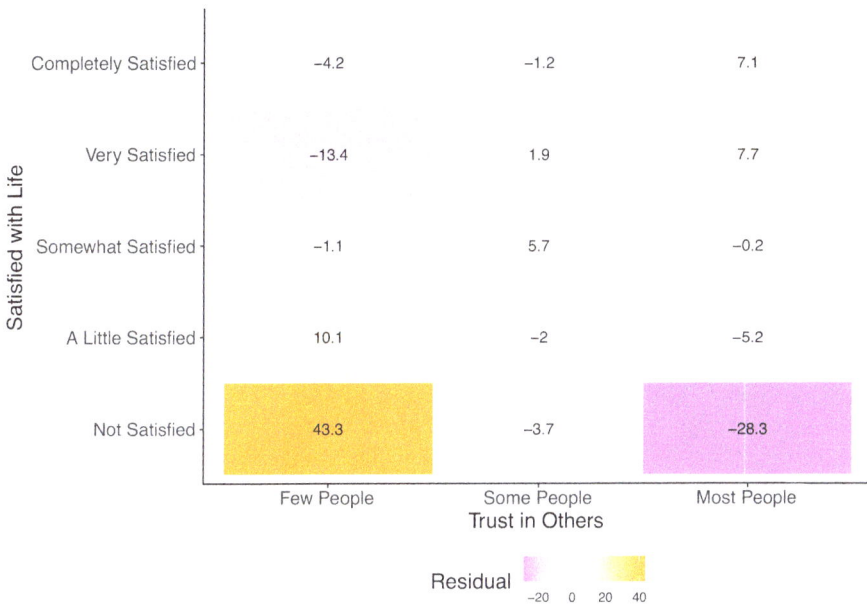

FIGURE 4.1
Heatmap of Chi-squared components from Table 4.4.

Log-Linear Models

As described in Chapter 3, the generalized linear model (GLM) is capable of predicting not only continuous responses but also limited dependent variables, such as counts. One popular approach to modeling contingency tables is the log-linear model, which models the count of respondents in each cell of a contingency table. We begin by describing the saturated model for the 2×2 table in Table 4.2. This model is presented in Eq. 4.1. The frequencies in the ith and j^{th} cell ($F_{i,j}$) are directly modeled as a function of the row margins, the column margins, and the interaction or association between them. The meaning of the coefficients will depend, of course, on how the independent variables are coded.[2] For these purposes, let us assume sex is dummy coded with female=1 and feeling safe walking alone at night is also dummy coded, with safe=1. The intercept (β_0) models the baseline count of respondents in the reference category, the effect of female (β_1) models how the row margins of Table 4.2 vary from the intercept, the effect of safe (β_2) models how the column margins vary from the intercept, and the interaction effect (β_3) models the unique contribution of the cell for females who feel safe.

$$F_{i,j} = \exp\left(\beta_0 + \beta_1 (Sex)_i + \beta_2 (Feel\ Safe)_j + \beta_3 (Sex \times Feel\ Safe)_{ij}\right) \quad (4.3)$$

In Eq. 4.3, we have inverted the link function in the expression so that the outcome is explicitly the observed counts. To model counts, we assume a Poisson distribution and implement a log link function. As such, the linear portion of the model is exponentiated to remove the log from the left-hand side of the equation. Modeling the log-counts is what characterizes this model as a "log-linear model." The counts are assumed to be linearly related to the row and column effects through the log transformation. This is also just a run-of-the-mill Poisson regression model, except that the outcome is the count of people in the cells of a contingency table rather than count-distributed individual variables (see Chapter 9).

To estimate Eq. 4.1 in R, we convert the contingency table into a data.frame object using the data.frame function. This reshapes the contingency table into analyzable data, as illustrated in Table 4.5. We then estimate the log-linear model with the following command:

```
glm(Count ~ Sex*Safe,data=lldat,family="poisson")
```

We use the glm command, which estimates the generalized linear model. Our outcome is the count of respondents in each cell, and we're modeling that as a function of both predictors and their interaction. The data are a reformatted cross-tabulation of the variables sex and satisfied. Finally, we use the "family=poisson" option. This option tells the glm function that we are assuming a Poisson distribution on our outcome variable. By default, the glm function uses a log link function for Poisson outcomes.

Model 1 in Table 4.6 is a *saturated model*, meaning that there are no degrees of freedom. The model simply reproduces the data. All of the cell counts in Table 4.2 can be reconstructed from Model 1. Given this model specification, by exponentiating the intercept, we reproduce the 396 female respondents who do not feel safe walking alone at night (i.e., exp(5.981)=395.84). Similarly, the 143 male respondents who do not feel safe can be recovered by subtracting the effect of male before exponentiating (i.e., exp(5.981−1.019)=142.88). The 803 females who feel safe walking alone at night (i.e., exp(5.981+.707)=802.72) and 851 males (i.e., exp(5.981−1.019+.707+1.077)=850.65) can be reproduced in the same fashion. As is implied by these expected counts, rows for male respondents see a 1.019 decrease in the expected log-frequency, columns for respondents who feel safe see a .707 increase in the expected log-frequency, and males who feel safe see a 1.077 increase in the expected log-frequency.

Model 2 in Table 4.6 is an *independence model* in the log-linear modeling tradition (Clogg and Eliason 1987; Knoke and Burke 1980). The model assumes that sex and feeling safe at night have independent effects on the cell counts but no interaction effect. This model predicts 295 females who report not feeling safe walking alone at night (exp(5.686)=294.7), 244 males

TABLE 4.5

Data Frame Format for the Contingency Table in 4.2

Sex	Safe	Count
Female	Not Safe	396
Male	Not Safe	143
Female	Safe	803
Male	Safe	851

TABLE 4.6

Summary of Two Log-Linear/Poisson Regression
Models Estimated on the Contingency Table in Table 4.2

	Model 1	Model 2
Male (M)	-1.019^{***}	-0.188^{***}
	(0.098)	(0.043)
Safe (S)	0.707^{***}	1.121^{***}
	(0.061)	(0.050)
M×S	1.077^{***}	
	(0.109)	
Constant	5.981^{***}	5.686^{***}
	(0.050)	(0.047)
Log-Likelihood	-15.866	-68.744

Note: $^{*}p<.05, ^{**}p<.01, ^{***}p<.001.$

who do not feel safe ($\exp(5.686-.188)=244.2$), 904 females who feel safe ($\exp(5.686+1.121)=904.2$), and 749 males who feel safe ($\exp(5.686-.188+1.121)=749.2$). More generally, cells for male respondents see a .188 decrease in the expected log-frequency, and cells for respondents who feel safe see a 1.121 increase in the expected log-frequencies.

Aside from interpreting the estimated regression coefficients, another means of interpreting log-linear models is to use the tools described above (odds and odds ratios) to examine the fitted counts. The odds that a male feels safe compared to not safe, according to the model, is 3.07 (i.e., $(749.2/(749.2+244.2))/(1-(749.2/(749.2+244.2)))=3.07$). Importantly, the odds that a female feels safe compared to not safe is also 3.07 (i.e., $(904.2/(904.2+294.7))/(1-(904.2/(904.2+294.7)))=3.07$). The *independence* model assumes that the cell counts are driven by the marginal distributions only, and that there is no association between sex and feeling safe walking alone at night. As such, men and women have the same odds of feeling safe. An implication of this is that odds ratios derived from the fitted values of independence models will always equal 1 (i.e., $3.07/3.07=1$).

Model 2 is *nested* in Model 1. Specifically, Model 2 constrains β_3 from Eq. 4.1 to be equal to zero. As discussed in the previous chapter, because the two models are nested, we can use the LR test of nested models. In this case, $G^2 = -2 \times (-68.74 - 15.866) = 105.8$. There is only one parameter differentiating Models 1 and 2, so the DF for this LR test is 1. The probability of observing the association we do in our contingency table is $<.001$ if the variables were independent. That is to say, there is sufficient empirical evidence to suggest that sex and life satisfaction are associated ($G^2_{(1)} = 105.8$, $p < .001$).

Below we present additional examples of log-linear models for contingency tables, but first we point out similarities between the Pearson Chi-squared test and the log-linear models in Table 4.6. The expected frequencies from the Pearson Chi-squared test are identical to the fitted values from the Independence Log-Linear model (Table 4.6, Model 2). For example, above we computed the expected count for females who do not feel safe from Model 2 to be 295. The same quantity is computed for the Pearson Chi-squared test (i.e., the sample of 2,193 is .547 female and .246 does not feel safe, so the expected count of females who do not feel safe is 295 ≈ 2,193(.547 × .246)). Indeed, the test statistics are asymptotically equivalent (Agresti 2003).

While the Pearson Chi-squared test and the log-linear models in Table 4.6 provide equivalent results, log-linear models are more flexible than the Chi-squared test of independence. Indeed, there is a robust and, in our view, underutilized literature on modeling contingency tables using log-linear models and their generalizations (Beller 2009; Breiger 1981; Clogg and Eliason 1987; Goodman 1979, 1981a, b; Haberman 1981). Below we review standard log-linear models as applied to contingency tables with more than two variables. To these data, we fit parameters capturing marginal effects and associations between sets of variables as interaction effects, as in Eq. 4.1. But log-linear and related models can go much further. There are many tailor-made models for contingency tables of mobility process (Goodman 1968; Sobel, Hout, and Duncan 1985) generalizations allowing analysts to control for continuous covariates (Logan 1983), or analysts may specify their own design matrices that explain patterns within the contingency table (e.g., Melamed 2015). These methods go beyond the scope of this book, but they are nonetheless important generalizations for the analysis of categorical data.

Our second example of log-linear models looks at the relationship between sex, education, and feeling safe walking alone at night. For education, we are concerned with four categories: high school or less, some college, a BA or BS degree, or those with graduate training. Table 4.7 presents the cross-tabulation of these variables using data from the ESS.

Table 4.8 presents a summary of four log-linear models that we fit to Table 4.7. Like R, our notation is inclusive of lower-order terms. So, for example, "Sex*Education" implies the interaction term and the main effects. In this example, we explicitly tested whether sex and education effects feeling safe at night, or if it can simply covary with sex and education. That is, we

TABLE 4.7

Cross-Tabulation of Sex, Education, and Feeling Safe Walking Alone at Night

Females	Feel Safe Walking Alone at Night	
Education	No	Yes
HS or Less	212	265
Some College	91	189
College Degree	22	80
Grad School	69	268
Males		
HS or Less	83	300
Some College	23	220
College Degree	11	93
Grad School	23	229

Note: Source is the European Social Survey, UK Sample.

treat safe as an outcome in our log-linear models. In this context, an interaction between safe and either other variable captures the effect of that variable on feeling safe (as our outcome). We did not explicitly adopt an inductive modeling strategy here; we do so in the following example. In this case, the LR test of nested models comparing Models 1 and 2 is not significant, meaning that constraining the terms in Model 1 to be equal to zero does not result in an overall loss of statistical information. Substantively, this means that sex and education do not need to covary or interact in their effect on feeling safe at night. We interpret this model in more detail below. Subsequent simplifications to the model specification, however, result in highly significant LR tests. We first tested whether sex covaries with feeling safe or if sex can be included as a simple main effect. Sex indeed effects the marginal distribution of feeling safe in Table 4.7 ($G^2_{(1)} = 105.76$, $p < .001$). Next, we test whether feeling safe at night varies with education or if education can be included as a simple main effect. This model implies that sex affects feeling safe at night but education does not. However, the LR test shows that education affects feeling safe ($G^2_{(3)} = 80.03$, $p < .001$).

Model 2 of Table 4.8 estimates ten parameters from the data. Fitting the marginals for education, sex, and safe takes five parameters (three for education, one for sex, and one for safe). The association between education and safe is three more and the association between sex and safe is one more. The intercept gives us the total of ten estimated parameters. Interpreting each is tedious. Instead, we often interpret odds or odds ratios from the model fitted values. To this end, Table 4.9 shows the marginal proportions of feeling safe at night by both sex and education. That both sex and education have only main effects shows up in these model predictions: Conditional on sex, for example, the effect of education is the same. The odds that a female with a

TABLE 4.8

Summary of Tests of Nested Models

Terms	No. of Parameters	LL	LR Tests
Model 1 Sex*Education*Safe	16	−50.80	
Model 2 Education*Safe+Sex*Safe	10	−56.45	Models 1 and 2, $G^2_{(6)}=11.31, p=.08$
Model 3 Education*Safe+Sex	9	−109.33	Models 2 and 3, $G^2_{(1)}=105.76, p<.001$
Model 4 Sex*Safe+Education	7	−96.47	Models 2 and 4, $G^2_{(3)}=80.03, p<.001$

graduate education feels safe at night, compared to not safe, is 2.8 times those same odds for females with a high school or less education (i.e., (.78/(1−.78))/ (.56/(1−.56))=2.79). Within rounding error, the same odds ratio is found for men: The odds that a male with a graduate education feels safe at night, compared to not safe, is 2.7 times those same odds for males with a high school or less education (i.e., (.91/(1−.91)) /(.79/(1−.79))=2.7). To examine the other effects, we look at odds ratios from the model predictions. As noted above, generally there are as many unique odds ratios as there is DF for the model.

Our last example of log-linear models shows its flexibility. We add whether the respondent is racially minoritized to our previous example. The contingency table is presented in Appendix A, as it has 32 rows (sex 2×education 4×minority 2×safe 2=32 cells). Table 4.10 summarizes ten different log-linear models that we fit to the four-way contingency table. Model 1 is saturated, with a four-way interaction and all lower-order terms. Model 2 constrains the four-way interaction but retains all three-way interactions. Comparing Models 1 and 2, the Akaike's Information Criteria (AIC) and Bayesian Information Criteria (BIC) are smaller for model 2, and the LR test of nested models is not significant. All the evidence leads us to prefer Model 2 to Model 1. Models 3–6 constrain one of the four three-way interaction terms in Model 2. Beginning with Model 3, we find an *increase* to the BIC and AIC when we constrain the three-way interaction between sex, being minoritized, and feeling safe. Furthermore, the LR test tells us that constraining this term results in a loss of statistical information. Models 4–6 all result in smaller or equivocal information criteria and none of the other LR tests are significant. This is evidence that the only three-way interaction warranted by the data is the one we constrained in Model 3 – the interaction between sex, being minoritized, and feeling safe.

We cannot constrain that single three-way interaction because the test of nested models between Models 2 and 3 was significant. Given this result, our next step is to test whether we can simultaneously remove the remaining three-way interactions. Model 7 includes the three-way interaction we

TABLE 4.9

Marginal Proportions of Feeling Safe Walking
Alone at Night from Model 2 Predicted Values

Females	Feel Safe	
Education	No	Yes
HS or Less	.44	.56
Some College	.30	.70
College Degree	.22	.78
Grad School	.22	.78
Males		
HS or Less	.21	.79
Some College	.12	.88
College Degree	.09	.91
Grad School	.09	.91

Note: Source is the European Social Survey.

could not constrain from above, and all two-way interactions. Model 7 yields smaller information criteria and an insignificant LR test, meaning that the empirical evidence prefers Model 7 to Model 2. Subsequent models in Table 4.10 systematically constrain each of the two-way interactions in Model 7. In all cases, the LR test is significant, meaning that we should not constrain those terms. As such, the preferred log-linear model for these data is Model 7, which can be interpreted using the same tools as those used above to understand the patterns or implications of the model.

Model 7 of Table 4.10 implies that men who are not minoritized are 3.4 times as likely as women who are not minoritized to report feeling safe at night, regardless of their educational attainment. Table A1 includes the fitted values from Model 7. With these, we can compute any odds ratio we wish. The odds a female who is not minoritized with high school or less education reports feeling safe, compared to not safe, is 1.21 (i.e., $(244.92/(244.92+201.6))$ / $(1-244.92/(244.92+201.6))=1.21)$, and those same odds for males is 4.09. Thus, the odds that a male who is not minoritized with high or less education will report feeling safe, compared to not safe, is 3.4 times those same odds for women (i.e., $4.09/1.21=3.4$). If we condition on any level of education, we will get this same odds ratio. This is because the three-way interaction between race, sex, and feeling safe does not covary with education in the model. Similarly, if we look at the odds ratios for minoritized individuals, we will get a different odds ratio, since the relationship between sex and safety varies by minority status in the model. The odds a female who is minoritized with high school or less education reports feeling safe, compared to not safe, is 1.16 (i.e., $(13.68/(13.68+11.8))$ / $(1-13.68/(13.68+11.8))=1.16)$, and those same odds for males is 1.55. Thus, the odds that a male who is minoritized with high or less education will report feeling safe, compared to not safe, is 1.33 times those same odds for women (i.e., $1.55/1.16=1.33$). Again, if we condition

TABLE 4.10

Model Summaries for Ten Log-Linear Models

Model	Terms	No. of Terms	LL	LR Test	BIC	AIC
1	E*M*F*S	32	−81.26		273.4	226.5
2	M*F*S+E*F*S+E*M*S +E*M*F	29	−81.63	Models 1 and 2, $G^2_{(3)}=.73, p=.87$	263.8	221.3
3	E*F*S+E*M*S+E*M*F	28	−85.03	Models 2 and 3, $G^2_{(1)}=6.81, p=.009$	267.1	226.1
4	M*F*S+E*M*S+E*M*F	26	−83.23	Models 2 and 4, $G^2_{(3)}=3.21, p=.36$	256.6	218.5
5	M*F*S+E*F*S+E*M*F	26	−82.98	Models 2 and 5, $G^2_{(3)}=2.71, p=.44$	256.1	218.0
6	M*F*S+E*F*S+E*M*S	26	−84.90	Models 2 and 6, $G^2_{(3)}=6.54, p=.08$	259.9	221.8
7	M*F*S+E*F+E*M+E*S	20	−87.98	Models 2 and 7, $G^2_{(9)}=12.70, p=.18$	245.3	216.0
8	M*F*S+E*M+E*S	17	−91.90	Models 7 and 8, $G^2_{(3)}=7.84, p=.049$	242.7	217.8
9	M*F*S+E*F+E*S	17	−99.47	Models 7 and 9, $G^2_{(3)}=22.99, p<.001$	257.9	232.9
10	M*F*S+E*F+E*M	17	−131.12	Models 7 and 10, $G^2_{(3)}=86.28, p<.001$	321.2	296.2

on any level of education, we will get the same odds ratio. Substantively, the model implies that the effect of sex is weaker for minoritized individuals. We also explore this relationship in more detail in Chapter 6.

Summary

In this chapter, we illustrated how to describe categorical variables in the R software environment, and we discussed two statistical models for categorical data analysis. Both models require exclusively categorical variables. The Pearson Chi-squared test evaluates whether two or more categorical variables are independent from one another by comparing the observed cell counts to expected cell counts. Log-linear models are an implementation of the general linear model that also models the cell counts of contingency tables. These models allow analysts to test which effects are warranted by the data in larger contingency tables.

In the next chapter, we show how logistic regression models combine elements of both linear regression and log-linear models. Like linear regressions, logistic regression allows analysts to model a certain type of categorical variable, namely, binary variables. Logistic regression also allows analysts to control for both categorical and continuous covariates. Like log-linear models, logistic regression uses a link function to ensure that model predictions are within the range of the outcome. Subsequent chapters address other specific outcome distributions, such as count (Chapter 9) or multinomial (Chapter 8) distributions.

Notes

1. The intuition of degrees of freedom can be straightforward. Consider a sample of four peoples' age, with the following data: 22, 34, 64, and 44. If we estimate one statistic from the data, say the mean of 41, we now have three degrees of freedom. Since we know the mean, any three other data points allow us to reconstruct the data.
2. Traditional log-linear models use effect coding rather than the more conventional dummy coding typically used in social science applications of regression models. Here, we use dummy coding for consistency with subsequent chapters and for purposes of comparability

Appendix A

TABLE A1

Cross-Tabulation of Sex, Education, Minority Status, and Feeling Safe Walking Alone at Night.

Education	Minority	Female	Safe	N	Model 7 Fitted Values
HS or Less	0	Female	0	200	201.6
Some College	0	Female	0	86	78.85
BA/BS	0	Female	0	21	21.49
Grad School	0	Female	0	56	61.07
HS Or Less	1	Female	0	10	11.8
Some College	1	Female	0	5	5.4
BA/BS	1	Female	0	1	2.02
Grad School	1	Female	0	12	8.78
HS Or Less	0	Male	0	75	70.37
Some College	0	Male	0	20	25.79
BA/BS	0	Male	0	8	7.85
Grad School	0	Male	0	17	15.99
HS Or Less	1	Male	0	8	9.23
Some College	1	Male	0	3	3.96
BA/BS	1	Male	0	3	1.65
Grad School	1	Male	0	6	5.16
HS Or Less	0	Female	1	243	244.92
Some College	0	Female	1	176	182.81
BA/BS	0	Female	1	76	71.13
Grad School	0	Female	1	237	233.15
HS Or Less	1	Female	1	19	13.68
Some College	1	Female	1	12	11.95
BA/BS	1	Female	1	3	6.37
Grad School	1	Female	1	30	32
HS Or Less	0	Male	1	287	288.12
Some College	0	Male	1	207	201.55
BA/BS	0	Male	1	83	87.54
Grad School	0	Male	1	206	205.79
HS Or Less	1	Male	1	12	14.28
Some College	1	Male	1	13	11.69
BA/BS	1	Male	1	10	6.96
Grad School	1	Male	1	23	25.06

Note: Source is the European Social Survey, UK Sample.

5

Regression for Binary Outcomes

In this chapter, we introduce binary regression models (BRMs), their relationship to simpler models overviewed in Chapters 3 and 4, and their derivation and interpretation. Although the BRM is not necessarily the foundation for other models considered in this book, the principles of fitting, testing, and interpreting these models carry over to every other model we cover. Accordingly, we dedicate two chapters to the BRM because they are simpler to illustrate the logic of these principles. To more clearly illustrate how these principles translate to more complex models, we use the same data set across these chapters and, to the extent possible, transformations and recodes of a common set of variables throughout. By "holding constant" the variables and data set used, we hope that relationships between different specifications and models become clearer to the reader.

As noted in Chapter 4, BRMs combine elements of both linear regression models overviewed in Chapter 3 and log-linear models covered in Chapter 4. We begin the chapter by reviewing the relationship between BRMs and these simpler models and motivate the move to an arguably more complex approach (although we do not believe BRMs are more complex to the applied researcher). We then derive the BRM using two complementary approaches. Both approaches lead to the same underlying statistical model, but one may be more theoretically or practically useful to your specific needs. We end this chapter with a basic overview of model presentation and interpretation. Chapter 6 dives into more advanced issues with interpretation.

Relationship to the Linear Probability and Log-Linear Models

Many social scientists are interested in binary outcomes. Using various predictors of substantive interest, demographers and health scholars may want to know whether someone has a particular health condition; political scientists may want to know whether someone votes for a particular candidate; and economists and sociologists may want to know whether a job candidate is hired. Each of these outcomes can only have one of two possible states. The outcome happened – the person has the health condition, the candidate is voted for, and the job candidate is hired – or it did not – the person does not have the health condition, the candidate was not voted for, and the job

DOI: 10.1201/9781003029847-5

candidate is not hired. As we discussed in Chapter 3, linear regressions are flexible and provide the best fitting line to describe the relationship between dependent and independent variables. Indeed, linear regressions are so flexible that they can and have been used to model binary outcomes such as these. The *linear probability model* (LPM) is an application of linear regressions to binary outcomes. By coding one state of the outcome as 0 and the other state of the outcome as 1, the LPM treats fitted values between 0 and 1 as probabilities that the outcome happens. In other words, the LPM extends OLS from:

$$y = Xb + \varepsilon \tag{5.1}$$

to

$$Pr(y_i = 1 \mid x_i) = Xb + \varepsilon \tag{5.2}$$

Despite its intuitive appeal, simpler interpretation, and select cases where it better fits the data (Timoneda 2021), the LPM can produce predicted probabilities below 0 and above 1. Accordingly, the model can produce estimates that are out of range and impossible to make sense of. There are at least two main ways in which scholars deal with the issue of out-of-range predictions. Each approach leads to the same statistical model. We begin by overviewing a "statistical approach" borrowing from log-linear models to use a link function to transform the predicted probabilities to a version that is constrained to be between 0 and 1. Then, we outline a "latent variable approach" to conceptualize the problem based on underlying propensities.

The Statistical Approach: Link Functions

The first major approach to generalize linear regressions to binary outcomes takes the LPM and constrains the predictions using a link function. To do this, we use the inverse of the cumulative distribution function (CDF) of common probability distributions as link functions. Because CDFs are between 0 and 1 by definition, they are useful constraints for our purposes. Two of the most common inverse CDF functions for binary outcomes is the *logit* link function – the inverse of the logistic CDF – and the *probit* link function – the inverse of the normal CDF. There are cases where the choice between logit and probit matters (Chen and Tsurumi 2010), but in general, they lead to virtually identical results and is a matter of personal preference and disciplinary convention

(Long 1997; Long and Freese 2006; Paap and Franses 2000). This is in part because the CDF of the logistic and normal distributions are so similar, as shown in Figure 5.1.

Our examples through this and other chapters are based on logit because of our own personal preferences and disciplinary conventions. Where the choice of a link function makes a difference in our discussions, we will indicate so; otherwise, the choice is inconsequential. Starting with the GLM described in Chapter 3:

$$\eta(y) = Xb \qquad (5.3)$$

η in this case is the logit. Because the logit is defined as the log of the odds of something happening, using the logit link transforms Eq. 5.3 into:

$$\ln\Omega\, Pr(y = 1 \mid x) = \ln\frac{Pr(x)}{1 - Pr(x)} \qquad (5.4)$$

This is why the logit is often referred to as the log-odds model and its interpretation can take the form of log-odds, a ratio of the odds, as well as the

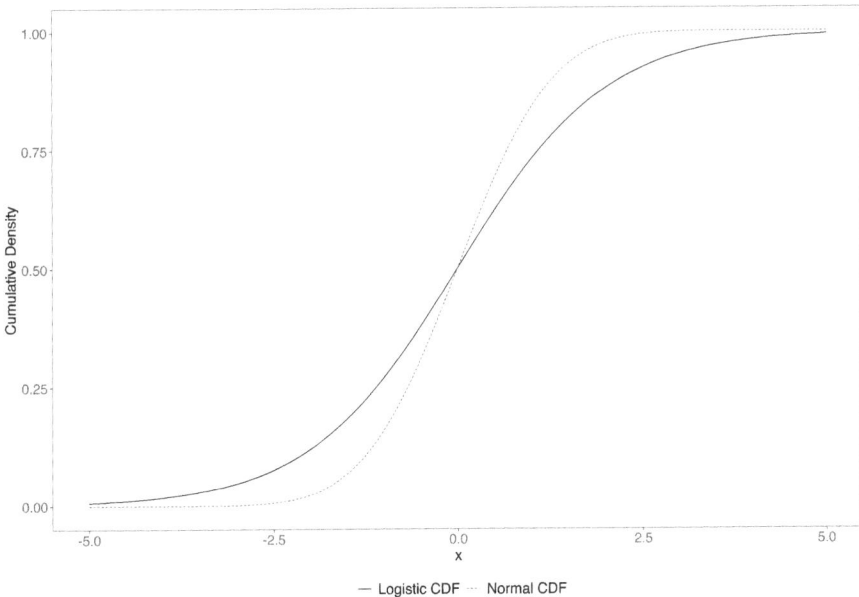

FIGURE 5.1
Cumulative density of logistic and normal CDFs.

predicted probability. If we were instead to use the probit link function, Eq. 5.3 is transformed into:

$$\phi^{-1}(y) = \phi^{-1}(Pr(x)) \qquad (5.5)$$

where ϕ^{-1} is the inverse normal CDF. In addition to logit and probit, R will allow the use of Cauchy CDF (Forbes et al. 2011) for binary outcomes. The default link function is the logit.

The Latent Variable Approach

Using the statistical approach implies that the BRM is a problem of constraining the linear regression to produce sensical results. An alternative approach leads to a mathematically equivalent model but conceptualizes the problem in more theoretical terms. Suppose that instead of viewing the BRM as a transformation of the linear regression, we view the linear regression as indicative of an underlying propensity and the binary outcome as the empirical manifestation of this underlying propensity. As Long and Freese (2006) posit, not all binary outcomes can be thought of in these terms. Nevertheless, for those that can, doing so provides a theoretically useful way to think of the problem and can justify a simpler interpretation of the methods discussed later in this chapter and in Chapter 6.

If we view the underlying propensity for the outcome as a latent or unobserved variable y^* that ranges from $-\infty$ to ∞, then this variable can be modeled using the standard linear regression model specified in Eq. 5.2. Using this approach, for each case, we treat the observed outcome as an indication of whether the latent variable is above or below some threshold τ. By convention, we talk about binary outcomes as successes and failures so we use $\tau=0$ as the threshold with successes being cases where y^* is positive and failures being cases where y^* is negative. In other words, the relationship between the observed outcome y and the unobserved propensity y^* is:

$$y_i = \begin{cases} 1 \text{ if } y_i^* > 0 \\ 0 \text{ if } y_i^* \leq 0 \end{cases} \qquad (5.6)$$

Note that the choice $\tau=0$ is arbitrary and does not affect the predicted probabilities because any other threshold is just a linear transformation of the latent variable. Also note that multiple τ can be specified, in which case we have the basis for the ordinal regression model discussed in Chapter 7.

Taking Eq. 5.2 and applying it to y^*, we run into the issue of what to do with the error term. Because y^* is unobserved, we cannot directly estimate it, but we also cannot ignore it because the probability is partially dependent on the distribution of the error term as shown in Figure 5.2. As shown in the figure, the error distribution affects whether and by how much the latent variable crosses the threshold τ. When comparing the relationship between the linear model y^* to the error term, we can see that it is mathematically tied to the linear prediction relative to the threshold. For observed cases closer to smaller or larger probabilities, only a small portion of the error distribution crosses the threshold in either direction, resulting in a small change in the predicted probabilities and smaller residuals. Closer to the threshold, proportionally more of the error distribution crosses the threshold, resulting in larger effects and larger residuals.

Despite the assumed variance affecting the spread and magnitude of regression coefficients, as proven in Long 1997 (pp. 49–50), the distribution of the error term does not affect the predicted probabilities themselves. Two commonly assumed distributions of the error term are $\text{Var}(\varepsilon)=1$ and $\text{Var}(\varepsilon)=\pi^2/3$, corresponding to the probit and logit models, respectively.

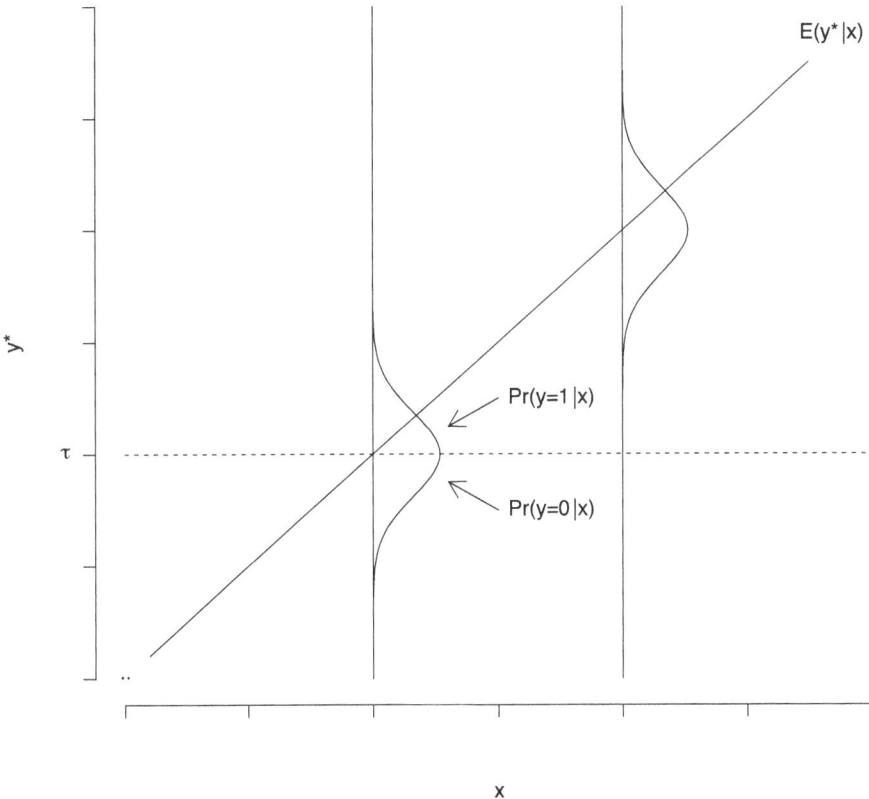

FIGURE 5.2
Relationship between y^*, $\text{Var}(\varepsilon)$, and $\Pr(y=1)$ in the latent variable approach.

Assuming a mean error of 0 and a variance of 1 results in a simpler form of the probit model:

$$Pr(y) = \int_{-\infty}^{Xb} \frac{1}{\sqrt{2\pi}} \, exp\left(-\frac{t^2}{2}\right) dt \qquad (5.7)$$

Similarly, assuming a mean error of 0 and variance of $\pi^2/3$ leads to a simpler form of the logit model:

$$Pr(y) = \frac{exp(Xb)}{1+exp(Xb)} \qquad (5.8)$$

Estimating a BRM in R

We can estimate a BRM similarly to how we would estimate a linear regression model using R's glm function. Again, we recommend creating a model object in your R workspace so that you can perform various post-estimation tasks with it. The call to estimate a binary logistic regression, the resulting output, and a prepared/formatted table of the output are as follows:

```
m1 <- glm(safe ~ religious + minority  + female + age + emp1 +
emp2,data=dat,family=binomial)
summary(m1)

Call:
glm(formula = safe ~ religious + minority + female + age + emp1 +
    emp2, family = binomial, data = dat)

Deviance Residuals:
    Min      1Q    Median       3Q      Max
-2.2630   0.4223   0.5756   0.8501   1.3142

Coefficients:
             Estimate Std. Error z value Pr(>|z|)
(Intercept)  1.604911   0.294423   5.451 5.01e-08 ***
religious   -0.018508   0.017782  -1.041 0.297942
minority    -0.168609   0.194665  -0.866 0.386408
female      -1.033938   0.112277  -9.209  < 2e-16 ***
age         -0.007041   0.003003  -2.344 0.019059 *
emp1         0.573451   0.263415   2.177 0.029482 *
emp2         1.016186   0.297467   3.416 0.000635 ***
---
Signif. codes:  0 '***' 0.001 '**' 0.01 '*' 0.05 '.' 0.1 ' ' 1
```

```
(Dispersion parameter for binomial family taken to be 1)

    Null deviance: 2406.5  on 2162  degrees of freedom
Residual deviance: 2279.8  on 2156  degrees of freedom
AIC: 2293.8

Number of Fisher Scoring iterations: 4
```

Walking through the call first, we see that similar to our `lm` call from Chapter 3, we specify a dependent variable – whether the respondent feels safe walking at night (`safe`) – followed by some independent variables – the respondent's religiosity (`religious`), whether they are a racial/ethnic minority (`minority`), whether they are female (`female`), their age in years (`age`), and whether they are traditionally employed (`emp1`) or self-employed (`emp2`) compared to unemployed (reference category). Unlike with the linear model, we specify a "binomial" family in our `glm` call to tell R that we want to run a BRM. By default, a binomial family model with no link implies a `logit` link. If we instead wanted to run a probit model, for example, we would issue the same call but specify `link= "probit"` as an option to the binomial family. The full call would be:

```
m1probit <- glm(safe ~ religious + minority  + female + age +
emp1 + emp2,data=dat,family=binomial(link="probit"))
```

TABLE 5.1

Summary of a Logistic Regression Model Predicting Whether the Respondent Feels Safe Walking Alone at Night

	−.019
Religious	(.018)
Minority (=1)	−.167
	(.195)
Female (=1)	−1.034***
	(.112)
Age	−.007*
	(.003)
Traditionally Employed[1]	.573*
	(.263)
Self-Employed	1.016*
	(.297)
Intercept	1.605***
	(.294)

Note: $*p < .05$, $**p < .01$, $***p < .001$. [1]Reference category is unemployed. Source is the European Social Survey, UK Sample.

The resulting output for the BRM also differs from the linear model in terms of the fit statistics reported – deviance residuals instead of residuals, null and residual deviance instead of an *F*-statistic on the model fit, and AIC instead of a multiple *R*-squared statistic. In addition, R reports the number of Fisher Scoring iterations, the number of tries it took to maximize the log likelihood.

Under the `Estimate` and `Std.Error` columns of the output, we have regression coefficients and their standard errors, respectively. We interpret these coefficients in substantive terms in the next section. For now, we focus on hypothesis tests and determining statistical significance. For individual coefficients, we can refer to the `z value` or its corresponding `Pr(>|z|)` columns to see whether these estimates exceed critical values determined by our respective hypotheses. Taking a look at the `female` variable, for example, we see that female respondents are less likely to feel safe at night compared to their male counterparts ($z = -9.21$, $p < .001$ for a two-tailed test). For more complex hypotheses involving multiple coefficients, however, we turn to Wald and likelihood-ratio (LR) tests.

Wald Tests

In its simplest form, the regression output we walked through includes a Wald test. The Wald test compares the maximum likelihood estimate of the coefficient to a null hypothesis considering variations in the estimates. The Wald test statistic under the null hypothesis asymptotically follows a chi-squared distribution. Under the special case of a single parameter test, the square root of a chi-squared test with 1 degree of freedom is equivalent to the *z* test reported in the regression output. We can illustrate this by performing a more general Wald test on each of our coefficients. For example, using the aod package (Lesnoff and Lancelot 2012), we can perform a Wald test on the `female` variable with the following call and resulting output:

```
wald.test(b = coef(m1), Sigma = vcov(m1), Terms = 4)

Wald test:
----------

Chi-squared test:
X2 = 84.8, df = 1, P(> X2) = 0.0
```

Calling the `wald.test` function with the model object `m1`, and specifying that we wanted to test the fourth coefficient in the model (`female`), produces a test statistic of 84.8, $p < .001$. $\sqrt{84.8} = 9.21 = z$ with the same *p*-value. Although illustrative, our use of the more general Wald test is for more complex hypotheses.

For example, suppose we wanted to know if employment significantly affects whether someone feels safe walking at night. We have two coefficients in the model and want to make a general statement about employment rather than traditional or self-employment. We can use a Wald test to jointly test whether the two coefficients are simultaneously equal to 0. To do this, we make the following call and obtain the following resulting output:

```
wald.test(b = coef(m1), Sigma = vcov(m1), Terms = 6:7)

Wald test:
----------

Chi-squared test:
X2 = 13.5, df = 2, P(> X2) = 0.0011
```

We can reject the null hypothesis that the effects of traditional and self-employment on fear for safety are simultaneously equal to 0 ($\chi^2_{(2)}=13.5$, $p<.01$).

LR Tests

An alternative way to conceptualizing hypothesis tests involves thinking about model fit. In the LR test, the null hypothesis is a restricted model where at least one coefficient is constrained to equal to zero. For more complex tests with two or more coefficients, the constraint is that these coefficients are simultaneously equal to 0. Thinking about the problem in these terms allows us to achieve the same goal by comparing the log likelihood of the full model to the log likelihood of the reduced model. Under the null hypothesis, the ratio of these log likelihoods also asymptotically follows a chi-squared distribution and allows us to test whether the full model significantly improves the log likelihood (Agresti 2003). As part of the `catregs` package, we use the `lr.test` function to test whether the two employment terms add to the model. First, we define a second model - one that constrains the effects of employment - and then we estimate the LR test.

```
m2<- glm(safe ~ religious + minority  + female +
age,data=dat,family=binomial)
lr.test(m2,m1)
   LL.Full  LL.Reduced  G2.LR.Statistic  DF  p.value
1 -1146.86   -1139.918          13.88333   2  0.00097
```

Performing the same hypothesis test using the LR test ($\chi^2_{(2)}=13.9$, $p<.001$) leads to a similar rejection of the null as with the Wald test. In fact, the Wald and LR tests are asymptotically equivalent. Although they are asymptotically equivalent, they can lead to different answers in finite samples. Because the Wald test was designed as an approximation to the LR test (Fox 1997), the LR test is preferred, especially if the computing time to estimate multiple models is negligible. The Wald test has the advantage of only requiring one run of the model compared to having to estimate both the full and reduced models to compare the log likelihoods.

Linear Combinations

In addition to hypotheses concerning the statistical significance of individual coefficients and the joint significance of groups of coefficients, we can also use the Wald and LR tests for linear combinations of coefficients. Take our employment coefficients. Suppose we want to test whether the effect of being traditionally employed is equivalent to the effect of being self-employed. In other words, we have competing theories that predict on the one hand, feeling unsafe at night is a by-product of the precarity of unemployment so being employed or not is what matters, not the type of employment. On the other hand, a competing theory argues that self-employment requires more independence, and that additional level of independence would lead to being less fearful.

We can perform a Wald test using the linearHypothesis function in the car package (Fox and Weisberg 2019) with the following call and resulting output:

```
linearHypothesis(m1, c("emp1 = emp2"))

Linear hypothesis test

Hypothesis:
emp1 - emp2 = 0

Model 1: restricted model
Model 2: safe ~ religious + minority + female + age + emp1 + emp2

  Res.Df Df  Chisq Pr(>Chisq)
1   2157
2   2156 1 7.8953    0.004956 **
---
Signif. codes:  0 '***' 0.001 '**' 0.01 '*' 0.05 '.' 0.1 ' ' 1
```

As is shown in the output, testing the hypothesis that emp1 = emp2 is equivalent to testing the hypothesis that emp1 – emp2 = 0. The resulting test provides enough evidence to reject the null hypothesis and conclude that traditional and self-employment have significantly different effects ($\chi^2_{(1)} = 7.9$, $p < .01$).

Performing the equivalent LR test requires that we specify the restricted model. Unlike the case where we are testing single or multiple coefficients, we are not simply leaving out the variables of interest to estimate the restricted model. What we are testing is that $\beta_{emp1} = \beta_{emp2}$ in the model:

$$
\begin{aligned}
\eta(y) = \alpha &+ \beta_{religious} \times religious + \beta_{minority} \times minority + \beta_{female} \times female \\
&+ \beta_{age} \times age + \beta_{emp1} \times emp1 + \beta_{emp2} \times emp2
\end{aligned}
\tag{5.9}
$$

if $\beta_{emp1} = \beta_{emp2}$, then the model can be written as:

$$\eta(y) = \alpha + \beta_{religious} \times religious + \beta_{minority} \times minority + \beta_{female} \times female$$
$$+ \beta_{age} \times age + \beta_{emp} \times emp1 + \beta_{emp} \times emp2 \quad (5.10)$$

where β_{emp} is the shared and equal coefficient for both the emp1 and emp2 variables. To estimate this restricted model, we make the call:

```
m3 <- glm(safe ~ religious + minority + female + age + I(emp1
+emp2),data=dat,family=binomial)
```

The I() function tells the glm function to estimate one coefficient for the sum of the emp1 and emp2 variables instead of estimating separate coefficients for these variables. Following this, we can perform the LR test as normal, resulting in the following output:

```
lr.test(m1,m3)

    LL.Full LL.Reduced G2.LR.Statistic DF p.value
1 -1139.918  -1144.116        8.396403  1 0.00376
```

As with our previous examples, the LR test leads to the same conclusion as the Wald test, with similar but slightly different test statistics, rejecting the null hypothesis that the effect of traditional and self-employment are equal ($\chi^2_{(1)} = 8.4, p < .01$).

Interpretation

Statistical significance gives us an indicator of how sure we are that relationships we observe in our models are due to chance, but it cannot tell us whether the relationship is substantively meaningful. For the BRM, there are three major ways one might go about interpreting regression results in substantive terms. We organize these approaches in order of least to most preferred generally, although individual project needs and disciplinary conventions will necessarily affect which approach is most preferred.

Regression Coefficients

Seemingly, the most straightforward way to interpret output from the BRM is to directly interpret the regression coefficients themselves. As with linear models, the BRM regression coefficients can tell us the direction of the variable's effect. Combined with tests of significance, we would be able to say if an independent variable has a significant effect and in which direction. However, the coefficients themselves cannot tell us the magnitude of the effect directly. Recall that our model is $\eta(y) = Xb$, not $y = Xb$. Therefore, the β coefficients reported in the regression output tell us the effects of the independent variables on a transformation of the dependent variable. This is not, more often than not, what we are interested in. Nonetheless, continuing our examples from above, the coefficient for the `female` variable would be interpreted along these lines:

> Being female is associated with a −1.03 change in the log-odds of feeling safe at night, all else constant.

What we really want to do is to speak to the effects of our independent variables on our outcome, feeling safe at night. Instead, the coefficients tell us about the log of the odds of feeling safe at night. This already cumbersome interpretation becomes worse with the probit, where we lose the log-odds interpretation and have to resort to interpreting the effect of the independent variable on the z-score of the probability of feeling safe at night.

A slightly more theoretically motivated way of interpreting the regression coefficients directly is to rely on the latent variable interpretation of the model. However, this still does not solve our problem of linking the latent propensity of the outcome (y^*) to the observed state of the outcome (y) in substantive terms. Furthermore, because y^* is dependent on the error variance we assumed, coefficient sizes are inherently tied to this assumption as well and requires standardizing the coefficients (Long and Freese 2006) or rescaling them (Breen, Karlson, and Holm 2013). This is especially problematic when comparing the magnitude of effects across groups (Long and Mustillo 2021) or across models (Winship and Mare 1983). Given these issues, we can speak to the direction and statistical significance of the coefficients using the latent variable interpretation, but that does not give us more traction on the problem than before.

Odds Ratios

A second way to interpret BRMs is in terms of changes in odds. Although it is theoretically possible to convert the probabilities from a probit into odds and interpret results from a probit in terms of changes in odds, it makes little sense to do so because it does not simplify any of the math relative to just calculating the predicted probabilities (it actually overcomplicates the problem). Therefore, interpretations based on odds are more naturally derived from the logit. To get from the log-odds to the more substantively meaningful odds, we can take the exponential of both sides of the equation:

$$e^{ln\Omega(y|x)} = e^{ln \frac{Pr\,(x)}{1-Pr\,(x)}} = e^{\alpha + \beta_1 x_1 + \beta_2 x_2 + \dots \beta_n x_n + \varepsilon} \tag{5.11}$$

When we do this, the additive linear model becomes a multiplicative model:

$$\Omega(y \mid x) = e^{\alpha} e^{\beta_1 x_1} e^{\beta_2 x_2} \dots e^{\beta_n x_n} e^{\varepsilon} \tag{5.12}$$

Suppose we wanted to examine a one unit change in variable x going from x to $x+1$. Focusing on x_1 and starting with the log-odds from the logit, we would predict the value of the outcome for $x+1$ and subtract it from the value of the outcome for x:

$$ln\Omega(y \mid x, x_1 + 1) - ln\Omega(y \mid x, x_1)$$
$$= \alpha + \beta_1(x_1 + 1) + \beta_2 x_2 + \dots \beta_n x_n + \varepsilon - [\alpha + \beta_1 x_1 + \beta_2 x_2 + \dots \beta_n x_n + \varepsilon] \tag{5.13}$$
$$= \alpha + \beta_1 + \beta_1 x_1 + \beta_2 x_2 + \dots \beta_n x_n + \varepsilon - [\alpha + \beta_1 x_1 + \beta_2 x_2 + \dots \beta_n x_n + \varepsilon]$$

In our exponentiated model, the subtraction becomes division, and this is how we arrive at an odds ratio when examining the changes in a logit coefficient holding all other variables constant:

$$\frac{\Omega(y \mid x, x_1 + 1)}{\Omega(y \mid x, x_1)} = \frac{e^{\alpha} e^{\beta_1} e^{\beta_1 x_1} e^{\beta_2 x_2} \dots e^{\beta_n x_n} e^{\varepsilon}}{e^{\alpha} e^{\beta_1 x_1} e^{\beta_2 x_2} \dots e^{\beta_n x_n} e^{\varepsilon}} = e^{\beta_1} \tag{5.14}$$

Therefore, for a unit change in a variable x, we can exponentiate the logit coefficient to interpret the odds ratio. Going back to the `female` variable example, we can interpret it as:

> Being female is associated with a decrease in the odds of feeling safe at night by a factor of .357 ($e^{-1.03}$ =.357), all else constant.

Because exponents are always positive, $e^x > 1$ would be interpreted as an increase in the odds, $e^x < 1$ would be interpreted as a decrease in the odds, and $e^x = 1$ would be no change. Odds less than one are harder to interpret than odds greater than one. Fortunately, we can invert the odds ratio and swap the reference category to make the interpretation more straightforward. For example, being male is associated with an increase in the odds of feeling safe at night by a factor of 2.8 (i.e., $1/\,e^{-1.03} = 2.8$). Note also that an odds ratio can be calculated for changes other than 1, but the change is multiplicative:

$$\frac{\Omega(y \mid x, x_k + \delta)}{\Omega(y \mid x, x_k)} = \frac{e^\alpha e^{\beta_1 x_1} \ldots e^{\delta \beta_k} e^{\beta_k x_k} \ldots e^{\beta_n x_n} e^\varepsilon}{e^\alpha e^{\beta_1 x_1} \ldots e^{\beta_k x_k} \ldots e^{\beta_n x_n} e^\varepsilon} = e^{\delta \beta_k} \qquad (5.15)$$

Therefore, although a 1 unit change in age would be associated with a decrease in the odds by a factor of .992 ($e^{-0.007}$ = .992), a decade change in age would be associated with a decrease in the odds by a factor of .932 ($e^{-0.007 \times 10}$ = .932).

Alternatively, it may be more intuitive to interpret the odds ratio as a percentage change in the odds instead of a factor change. The percentage change in the odds can be calculated by:

$$percentage\ change\ in\ odds = 100 \times \left(e^{\delta \beta_k} - 1\right) \qquad (5.16)$$

Our interpretation based on the percentage change in odds would be that a decade increase in age is associated with a 6.8% decrease in the odds of feeling safe at night ($100 \times (e^{-0.007 \times 10} - 1) = -6.8\%$). We can get the odds ratio, corresponding confidence interval, and the percentage change in the odds using the list.coef function from the catregs package and make the following call to receive the following output:

```
list.coef(m1)
```

$out

variables	b	SE	z	ll	ul	p.val	exp.b	ll.exp.b	ul.exp.b	percent	CI
1 (Intercept)	1.605	0.294	5.451	1.028	2.182	0.000	4.977	2.795	8.864	397.742	95%
2 religious	-0.019	0.018	-1.041	-0.053	0.016	0.232	0.982	0.948	1.016	-1.834	95%
3 minority	-0.169	0.195	-0.866	-0.550	0.213	0.274	0.845	0.577	1.237	-15.516	95%
4 female	-1.034	0.112	-9.209	-1.254	-0.814	0.000	0.356	0.285	0.443	-64.440	95%
5 age	-0.007	0.003	-2.344	-0.013	-0.001	0.026	0.993	0.987	0.999	-0.702	95%
6 emp1	0.573	0.263	2.177	0.057	1.090	0.037	1.774	1.059	2.973	77.438	95%
7 emp2	1.016	0.297	3.416	0.433	1.599	0.001	2.763	1.542	4.949	176.264	95%

The list.coef function reports the regression coefficient, standard error, and various indicators of statistical significance in the first six columns of output. The odds ratio is reported in the exp.b column. Note that because we are returning the full model, R produces the odds ratio for the intercept as well as our independent variables, but it does not make sense to interpret the intercept because it does not change. The other noteworthy point from this output is the asymmetrical confidence intervals. This is because exponents are a nonlinear transformation. Therefore, when reporting odds ratios, we tend to report confidence intervals instead of exponentiated standard errors. The `percent` column reports the percentage change in odds as calculated in Eq. 5.16. Finally, the `CI` column tells us the level at which the confidence intervals are calculated. In this example, the `ll`, `ul`, `ll.exp.b`, and `ul.exp.b` columns report the lower and upper bounds of a 95% confidence interval.

The output of `list.coef` may be plotted to create coefficient plots. Figure 5.3 presents a coefficient plot for the odds ratios from Model 1. The plot illustrates the odds ratios and their 95% confidence intervals. We find coefficient plots to be an effective alternative means to tables for summarizing statistical models, particularly for presentations.

Another quirk with the odds ratio and its nonlinearity comes from comparing them. An odds ratio of 2 is equivalent in magnitude to an odds ratio of 0.5 (or ½) despite being a 1-point versus a 0.5-point difference from the baseline odds ratio of 1. More importantly, the same odds ratio can result from very different changes in probabilities. Suppose that an event is very rare and only has a probability of 0.001, or 1 to 1,000 odds. A variable that

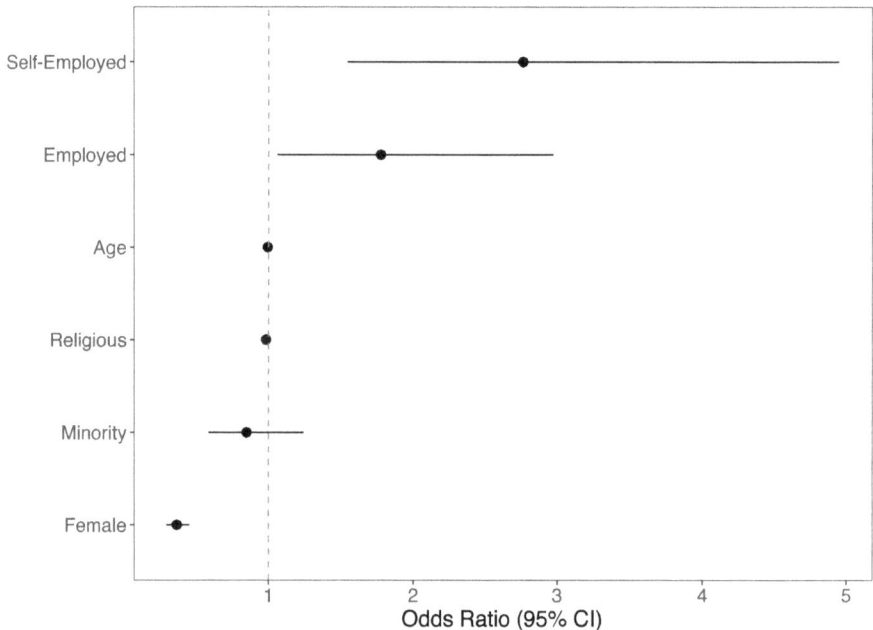

FIGURE 5.3
Coefficient plot illustrating odds ratios and corresponding confidence intervals from Model 1.

doubles the odds (OR=2) would increase the probability to 0.002, changing the probability by 0.001. A variable that doubles the odds of a common event with 1 to 1 odds would increase the probability from 0.5 to 0.667, changing the probability by 0.167. Substantively, a 0.167 change in probability is much larger than the 0.001 change, but both have the same odds ratio because of the different baseline probability from which we started. In other words, the odds ratio is a more meaningful way of interpreting the logistic regression than the raw coefficients, but they still ultimately depend on the predicted probabilities from our model.

Predicted Probabilities

The last and preferred way of interpreting BRMs involves changes to the predicted probabilities of the outcome. Compared to regression coefficients and odds ratios, predicted probabilities are substantively most interpretable. Rather than discussing the odds, log of the odds, or the *z*-score of the predicted probabilities, we are directly discussing the predicted probability of the outcome based on our model. It is worth noting that the estimates described in this section are often called *marginal effects at means* (MEM; Long and Freese 2006; Mize 2019). *Marginal effects at representative values* can be implemented by setting specific values (other than means) in the design matrix. The `catregs` package includes four functions to generate and test for differences in predicted probabilities; that is, to compute marginal effects at means or marginal effects at representative values. The four functions include:

```
margins.des
margins.dat
first.diff.fitted
second.diff.fitted
```

In order to generate predicted probabilities, we need to input a design matrix to tell R at which level of the independent variables to calculate predictions. The `margins.des` function creates this design matrix for `margins.dat` to calculate the predictions. Continuing our example using the `female` variable, we make the following call to create the design matrix:

```
design <- margins.des(m1,ivs=expand.grid(female=c(0,1)))
```

The function takes a model object and the levels of independent variables for which you want to generate predicted probabilities. The expand.grid function ensures a factorial-crossing of all variables listed in `ivs`. Any variable not specified in `ivs` will be set at its respective mean. Note that this implies that you should exclude factor variables from being included in the design matrix using the `exc` option; these will be proportionally weighted by `margins.dat`. Returning to the example above, we are creating a design matrix that says to generate predicted probabilities for respondents who are 0 and 1 (male and female, respectively) on the female variable setting the other variables in our model at their mean values. We then give the margins.dat function this design matrix using the following call, resulting in the following output:

```
margins.dat(m1,design)
```

```
  female religious minority    age empl emp2 fitted    se   ll    ul
1      0     3.602    0.077 53.146 0.799 0.168  0.856 0.011 0.833 0.878
2      1     3.602    0.077 53.146 0.799 0.168  0.678 0.014 0.651 0.706
```

The fitted column tells us the predicted probability from the model (conditional on the covariates earlier in the row); the se tells us the Delta method standard error for that prediction, with the next two columns being the corresponding lower and upper limits of a confidence interval. The first several columns in the output tells us the values at which the prediction is estimated (3.602 for the religious variable, 0.077 for minority, 53.146 for age, 0.799 for empl, and 0.168 for emp2). One may interpret this output along these lines:

> Male respondents have a predicted probability of feeling safe at night of 0.856 compared to a predicted probability of 0.678 for female respondents, holding other variables at their means.

We are not limited to categorical comparisons using margins.dat. For example, changing the design matrix call to examine 5-year increments of age instead of respondent sex, we can make the following call, resulting in the following output:

```
design <- margins.des(m1,expand_grid(age=seq(20,80,5)))
pdat <- margins.dat(m1,design)
pdat
```

```
   age religious minority female empl  emp2 fitted    se   ll    ul
1   20     3.602    0.077  0.544 0.799 0.168  0.810 0.018 0.775 0.845
2   25     3.602    0.077  0.544 0.799 0.168  0.805 0.016 0.773 0.836
3   30     3.602    0.077  0.544 0.799 0.168  0.799 0.015 0.771 0.827
4   35     3.602    0.077  0.544 0.799 0.168  0.793 0.013 0.768 0.819
5   40     3.602    0.077  0.544 0.799 0.168  0.787 0.011 0.765 0.810
6   45     3.602    0.077  0.544 0.799 0.168  0.781 0.010 0.761 0.802
7   50     3.602    0.077  0.544 0.799 0.168  0.775 0.010 0.757 0.794
8   55     3.602    0.077  0.544 0.799 0.168  0.769 0.010 0.750 0.788
9   60     3.602    0.077  0.544 0.799 0.168  0.763 0.010 0.743 0.783
10  65     3.602    0.077  0.544 0.799 0.168  0.757 0.012 0.734 0.779
11  70     3.602    0.077  0.544 0.799 0.168  0.750 0.014 0.723 0.776
12  75     3.602    0.077  0.544 0.799 0.168  0.743 0.016 0.712 0.774
13  80     3.602    0.077  0.544 0.799 0.168  0.737 0.018 0.701 0.773
```

Although it was fairly intuitive to interpret the raw predicted probability of male compared to female respondents, when evaluating across a range of a variable like age, it might make more sense to plot the predicted probabilities. Giving the object returned by `margins.dat` to a plotting function like `ggplot`, we can create a predicted probability plot like Figure 5.4.

Of course, we are often interested in whether the differences between various values of predicted probabilities are significantly different from each other in addition to what the values are themselves. To do this, we can use the `first.diff.fitted` function (and for more advance applications covered in Chapter 6, `second.diff.fitted`). Going back to our respondent sex example, we can make the following call to test how much female and male respondents differ in their predicted probability of feeling safe at night:

```
first.diff.fitted(m1, design, compare=c(1,2))

    term    est std.error statistic p.value      ll      ul
1    b1 0.177     0.018     9.834        0 0.142 0.213
```

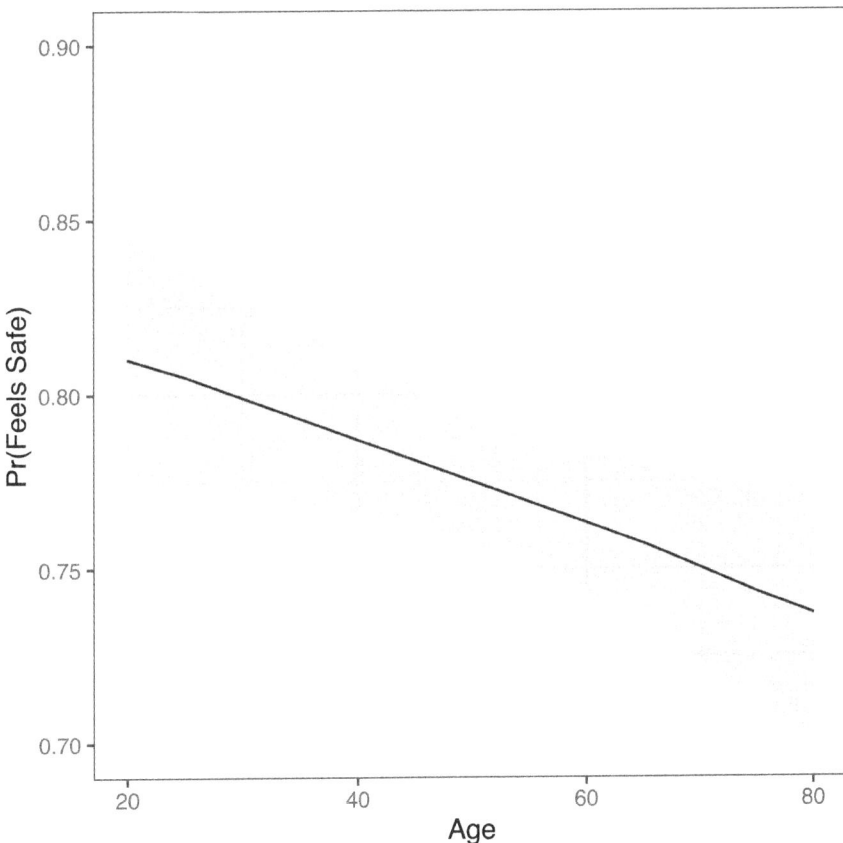

FIGURE 5.4
Predicted probability of feeling safe by age.

Walking through the function, `first.diff.fitted` takes the model object and design matrix created by `margins.des` as input. The `compare` option specifies which rows of the design matrix to compare. In this case, `compare=c(1,2)` tells us $Pr\left(female=0,\, x_k=\overline{x_k}\right)-Pr(y=1\,|\,female=1,\, x_k=\overline{x_k})$. In other words, it tells us how much more likely male respondents are to feel safe than female respondents. We could easily reverse the option to be `compare=c(2,1)` to get how much *less* likely female respondents are to feel safe than male respondents. Interpreting this output in substantive terms, we would say something along the lines of:

> Male respondents have a 0.177 higher predicted probability (0.856 vs. 0.678) of feeling safe at night compared to female respondents, holding other variables at their means ($p<.001$, two-tailed).

We can also run the `first.diff.fitted` function with our larger predicted probabilities table based on age. To compare how much more likely a 30-year-old is to feel safe at night relative to a 65-year-old, for example, we would run the following call based on our age design matrix, which results in the following:

```
first.diff.fitted(m1, design, compare=c(3,10))

   term   est std.error statistic p.value    ll    ul
1    b1 0.042     0.018     2.407   0.016 0.008 0.077
```

In this case, the prediction for 30-year-olds is the third row in our design matrix and the prediction for 65-year-olds is the tenth row. The difference in predicted probabilities across these two age values is 0.042, $p<.05$.

Average Marginal Effects

Rather than setting covariates to their means or to representative values, another approach is to compute the average marginal effect (AME) for a unit change in a predictor. The AME uses the observed data and computes the marginal effect in the outcome as a focal variable changes by one unit (Searle, Speed, and Milliken 1980; Wooldridge 2010). The margins package (Leeper 2021) includes the margins function that computes AMEs and conditional AMEs (discussed in the next chapter).

Consider, for example, the AME of female in Model 1. This refers to $Pr\left(female=0,\, x_k=\overline{x_{ki}}\right)-Pr(y=1\,|\,female=1,\, x_k=\overline{x_{ki}})$, and can be estimated with the following call:

```
> summary(marginaleffects(m1))
       Term    Effect Std. Error z value   Pr(>|z|)      2.5 %     97.5 %
1 religious -0.003224   0.003095 -1.0418 0.29748258 -0.009290  0.0028414
2  minority -0.029373   0.033894 -0.8666 0.38614600 -0.095803  0.0370571
```

```
3     female -0.180125   0.018536 -9.7176 < 2.22e-16 -0.216454 -0.1437949
4        age -0.001227   0.000520 -2.3588 0.01833189 -0.002246 -0.0002074
5       emp1  0.099899   0.045710  2.1855 0.02885247  0.010309  0.1894882
6       emp2  0.177024   0.051391  3.4447 0.00057175  0.076300  0.2777478
```

Note that this returns the AME for every predictor variable. If we wanted to restrict it to estimate just the AME for females (e.g., if the model was complicated and AMEs took a while to estimate), the code would be `marginaleff ects(m1,variables="female")`. The AME for female is –.18. Interpreting this in substantive terms:

> Male respondents have a 0.18 higher predicted probability of feeling safe at night compared to female respondents, holding other variables at their observed values ($p < .001$, two-tailed).

Note that the AME (–.1801) is similar to the MEM (–.177) reported above (the `first.diff.fitted` output), as we would hope. The former is based on the observed data and the latter is based on holding covariates at their observed means. In the next chapter, on more advanced topics, we illustrate how to probe statistical interactions using MEMs and AMEs. For now, we turn to case-level model diagnostics.

Diagnostics

Before moving onto more complex issues of interpretation, it is important to perform some basic model diagnostics. Looking at residuals, outliers, and influential cases can be useful for alerting us to potential issues with our models. Much of the logic of model diagnostics for BRMs are similar to those for linear models (Belsley, Kuh, and Welsch 2005; Fox 1997). However, as foreshadowed in our initial walkthrough of the output from the logistic regression above, the specific measures are slightly different. We walk through these below.

Residuals

In the GLM there are often multiple versions of model residuals. Various forms of residuals are measures of by how much individual observations differ from their predicted values. The *Pearson residual* is calculated by dividing the observation's residual by its standard deviation:

$$r_i = \frac{y_i - \hat{\pi}_i}{\sqrt{\hat{\pi}_i \left(1 - \hat{\pi}_i\right)}} \tag{5.17}$$

where $\hat{\pi}_i$ is the predicted probability from our model for the ith observation. Because the variance of the error for a BRM is heteroskedastic, we may wish to standardize the Pearson residual to take into account the correlation between y and $\hat{\pi}$ (Hardin and Hilbe 2007). The *standardized Pearson residual* is calculated by:

$$r_i^{std} = \frac{r_i}{\sqrt{1-h_{ii}}} \tag{5.18}$$

where $\qquad\qquad h_{ii} = \hat{\pi}_i\left(1-\hat{\pi}_i\right)x_i\widehat{Var}\left(\hat{\beta}\right)x_i'$

A third form of the residual looks at the contribution of each observation to the log likelihood of the model. This is called the *deviance residual*, calculated as:

$$d_i = s_i\sqrt{-2\left[y_i\log\hat{\pi}_i + \left(1-y_i\right)\log\left(1-\hat{\pi}_i\right)\right]} \tag{5.19}$$

where $s_i = 1$ if $y_i = 1$ and $s_i = -1$ if $y_i = 0$

The `diagn` function in `catregs` computes the relevant diagnostics for each class of models discussed in this book. For example, the following call calculates and displays the first ten observations' residuals:

```
> diags <- diagn(m1)
> diags[1:10,c(1,3,6)]
```

	pearsonres	stdpres	devres
1	-1.1621818	-1.1728961	-1.3074394
2	0.4025085	0.4027731	0.5480036
3	0.8175877	0.8246206	1.0118249
4	0.7344900	0.7350393	0.9289153
5	0.7122279	0.7126231	0.9058687
6	-2.4361291	-2.4429361	-1.9680149
7	0.3602473	0.3607942	0.4940066
8	0.8272186	0.8288699	1.0211215
9	0.3896022	0.3898790	0.5316337
10	0.4555187	0.4560467	0.6140845

As shown in the example, the Pearson and standardized Pearson residuals are fairly similar, but we should prefer the standardized version for its constant variance. Having calculated these residuals, we can create a residual plot to identify large residuals. Unfortunately, there is not an agreed upon standard for evaluating whether a given residual is too large. However, plotting the residuals can highlight cases that seem to "stand out," perhaps indicating errors in the data or at the very least cases worthy of a double check. Figure 5.5 does not appear to show any particularly peculiar cases.

Alternatively, we can identify cases with extreme values based on our preferred residual measure and see if our conclusions are affected. The following call selects out cases where the absolute value of the deviance residual is greater than 2 and compares a model without those cases to our original model:

```
> dat.5 <- data.frame(dat,diags)
> m1.sr <- glm(safe ~ religious + minority  + female + age +
emp1 + emp2,data=filter(dat.5,devres >= -2 & devres
<=2),family="binomial")
>round(cbind(full.coef=coef(m1),sr.coef=coef(m1.sr),full.
se=sqrt(diag(vcov(m1))),sr.se=sqrt(diag(vcov(m1.sr))),
+         full.z=coef(m1)/sqrt(diag(vcov(m1))),sr.z=coef(m1.sr)/
sqrt(diag(vcov(m1.sr)))),3)
```

	full.coef	sr.coef	full.se	sr.se	full.z	sr.z
(Intercept)	1.605	1.997	0.294	0.307	5.451	6.510
religious	-0.019	-0.036	0.018	0.019	-1.041	-1.932
minority	-0.169	-0.153	0.195	0.204	-0.866	-0.751
female	-1.034	-1.335	0.112	0.124	-9.209	-10.802
age	-0.007	-0.010	0.003	0.003	-2.344	-3.025
emp1	0.573	0.654	0.263	0.270	2.177	2.423
emp2	1.016	1.486	0.297	0.318	3.416	4.674

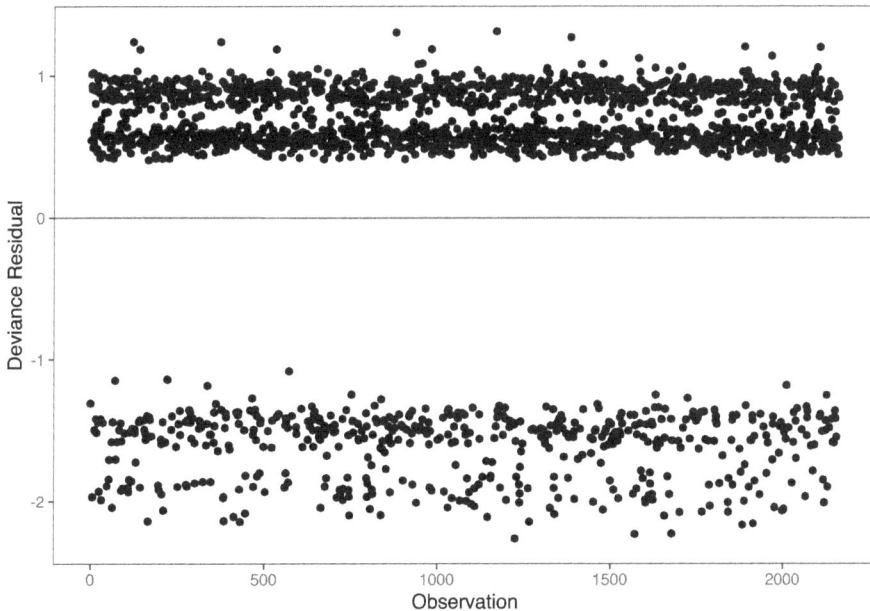

FIGURE 5.5
Deviance residual plot.

As shown in the output, the direction and significance of our coefficients are unaffected by these cases. The magnitude is also similar with and without these cases.

Influential Cases

Cases with large residuals may not necessarily have strong influences on our model's parameters. Another often used diagnostic is Cook's D, which is defined as the difference between the model's parameters with and without a given observation. If the removal of an observation greatly changes the model's parameters, it is an influential case and is worthy of further investigation. Cook's D is calculated as:

$$D_i = \frac{\sum_{j=1}^{n}\left(\hat{Y}_j - \hat{Y}_{j(i)}\right)^2}{(p+1)\hat{\sigma}^2} \qquad (5.20)$$

The `diagn` function includes Pregibon's (1981) Delta-Beta measure of influence, which is an approximation of Cook's D based on the standardized Pearson's residual from Eqs. 5.17 and 5.18:

$$\Delta\hat{\beta}_i = \frac{r_i^2 h_{ii}}{\left(1 - h_{ii}\right)^2} \qquad (5.21)$$

As with residuals, how influential an observation is to be considered problematic is more an art than a science. We can adapt many of the same tools of plotting, identifying, and respecifying the model to account for influential cases. Figure 5.6, for example, shows several influential cases worthy of further investigation. A general rule of thumb for a Cook's D/Delta Beta that is too large is $4/n$ where n is the number of observations. The red line identifies this cut point in the figure. Using this rule of thumb, we can print out influential cases and re-estimate models excluding them to see how these cases affect our conclusions.

The following call re-estimates and displays a comparison of our model with and without the influential cases:

```
> m1.cd <- glm(safe ~ religious + minority  + female + age +
emp1 + emp2,data=filter(dat.5,deltabeta<=(4/nrow(diagn))),
family="binomial")
Warning message:
glm.fit: fitted probabilities numerically 0 or 1 occurred
> round(cbind(full.coef=coef(m1),cd.coef=coef(m1.cd),full.
se=sqrt(diag(vcov(m1)))),cd.se=sqrt(diag(vcov(m1.cd)))),
+        full.z=coef(m1)/sqrt(diag(vcov(m1))),cd.z=coef(m1.cd)/
sqrt(diag(vcov(m1.cd)))),3)
```

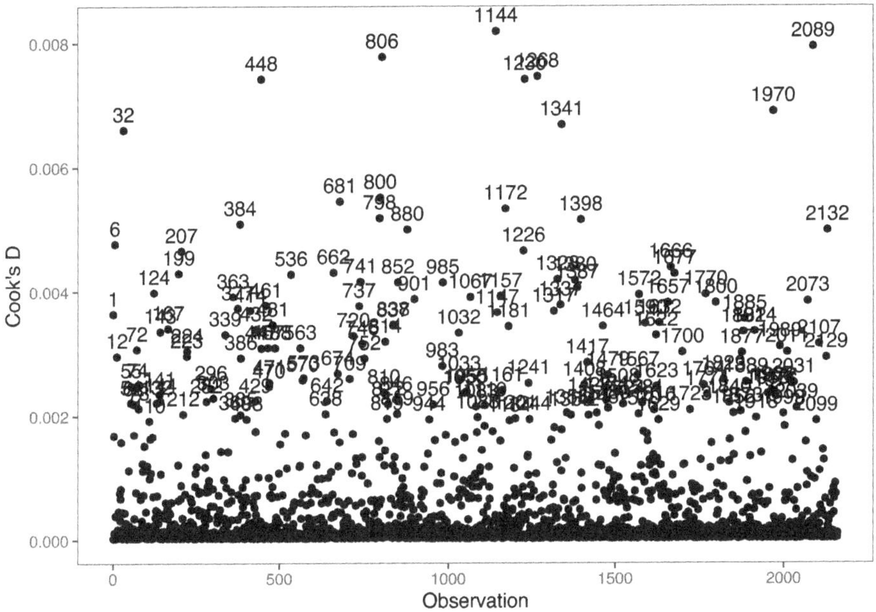

FIGURE 5.6
Cook's D plot.

	full.coef	cd.coef	full.se	cd.se	full.z	cd.z
(Intercept)	1.605	19.322	0.294	1180.172	5.451	0.016
religious	-0.019	-0.014	0.018	0.022	-1.041	-0.633
minority	-0.169	16.980	0.195	557.163	-0.866	0.030
female	-1.034	-1.361	0.112	0.143	-9.209	-9.521
age	-0.007	-0.013	0.003	0.004	-2.344	-3.496
emp1	0.573	-16.536	0.263	1180.172	2.177	-0.014
emp2	1.016	0.527	0.297	1231.925	3.416	0.000

Here, we can see that the model removing these case results is nonsensically large coefficients and standard errors, suggesting that we may have overcorrected. In examining Figure 5.6, we can see a clear delineation about $D = 0.006$ where nine cases are drastically different from the other cases. Using $D = 0.006$ as the threshold results in a more similar model to our original model:

```
> m1.cd2 <- glm(safe ~ religious + minority  + female + age +
emp1 + emp2,data=filter(dat.5,deltabeta<=.006),family=
"binomial")
> round(cbind(full.coef=coef(m1),cd.coef=coef(m1.cd2),full.
se=sqrt(diag(vcov(m1))),cd.se=sqrt(diag(vcov(m1.cd2))),
+             full.z=coef(m1)/sqrt(diag(vcov(m1))),cd.
z=coef(m1.cd2)/sqrt(diag(vcov(m1.cd2)))),3)
          full.coef cd.coef full.se cd.se full.z    cd.z
```

```
(Intercept)      1.605    1.924    0.294 0.313    5.451    6.141
religious       -0.019   -0.021    0.018 0.018   -1.041   -1.195
minority        -0.169   -0.046    0.195 0.202   -0.866   -0.230
female          -1.034   -1.096    0.112 0.114   -9.209   -9.582
age             -0.007   -0.007    0.003 0.003   -2.344   -2.414
emp1             0.573    0.315    0.263 0.279    2.177    1.130
emp2             1.016    0.829    0.297 0.314    3.416    2.643
```

Measures of Fit

Model diagnostics can warn us of potential problems within our model, but they are less useful across models. What if we wanted to compare competing model specifications? Here, we turn to measures of model fit to provide guidance. As an initial measure of fit, R already reports the "Null deviance" and "Residual deviance" for our model in its glm output:

```
    Null deviance: 2406.5  on 2162  degrees of freedom
Residual deviance: 2279.8  on 2156  degrees of freedom
```

These deviance measures compare our model to an intercept-only model (the null deviance). What we ideally want to see is a reduction in the deviance, which tells us we are doing better than just guessing the grand mean for every respondent. A more formal way of testing that we are improving the fit is to perform an LR test comparing our model to the null model:

```
m0<-glm(safe ~ 1,data=dat,family=binomial)
lr.test(m1,m0)

     LL.Full LL.Reduced G2.LR.Statistic DF p.value
1 -1139.918  -1203.251         126.6658  6       0
```

As we can see, our model significantly improves the log likelihood ($\chi(6)2 = 126.67$, $p < .001$). We can also use the pR2, AIC, and BIC functions to obtain various pseudo-R^2 measures (see Long 1997 for an overview of these) as well as the Akaike Information Criterion and the Bayesian Information Criterion:

```
pR2(m1)
fitting null model for pseudo-r2
         llh          llhNull            G2        McFadden
r2ML             r2CU
-1.139918e+03 -1.203251e+03  1.266658e+02  5.263480e-02
5.687856e-02  8.473037e-02

AIC(m1)
[1] 2293.837

BIC(m1)
```

We interpret the pseudo-R^2 measures similarly to how we interpret R^2 in a linear regression. The fourth through sixth columns report McFadden's pseudo-R^2, the maximum likelihood pseudo-R^2, and Cragg and Uhler's pseudo-R^2, respectively. As shown in the output, they are similar to one another. We do not advocate for the use of any one of these in particular, and agree with Long and Freese's (2006:221) sentiment, "there is no convincing evidence that selecting a model that maximizes the value of a pseudo-R^2 results in a model that is optimal in any sense other than the model has a larger value of that measure."

In contrast, selecting models based on information criteria like AIC and BIC has a stronger theoretical foundation (Kuha 2004; see Chapter 3). Both AIC and BIC quantify the trade-off between improving likelihood and model parsimony. To use either measure, we would need to compare the AIC and BIC across models, with lower values indicating better fit (higher maximized likelihoods relative to the number of parameters used). Revisiting our Model 2 from earlier in the chapter where we do not account for employment status, we can generate and compare the AIC and BIC for this more parsimonious model to our full model to see if it is a better fitting model:

```
AIC(m2,m1)
BIC(m2,m1)

     df       AIC
m2   5  2303.720
m1   7  2293.837

     df       BIC
m2   5  2332.116
m1   7  2333.591
```

Here, we can see how the different treatment of the number of parameters can lead to differing conclusions. The difference between Models 1 and 2 is the inclusion of the two (statistically significant) indicator variables for traditional and self-employment. This addition, based on our LR test from earlier, is justified because it improves the likelihood of the model significantly. The AIC yields the same conclusion, with Model 1 showing a better comparative fit to Model 2 (AIC for Model 1 of 2293.8 is less than AIC for Model 2 of 2303.7). However, the BIC indicates that Model 2 might better fit the data (BIC for Model 2 of 2332.1 is less than the BIC for Model 1 of 2333.6). Using Raftery's (1995) guidelines for the BIC, the 1.475 difference in BIC scores provides "weak evidence" that Model 2 is a better fitting model. Given conflicting evidence (the additional variables are predictive and improves the fit, but does it improve enough to justify adding two more coefficients to the model?), which model is better depends on the theoretical and practical importance of taking into account employment.

6

Regression for Binary Outcomes – Moderation and Squared Terms

Building on the previous chapter, in this chapter, we describe how to assess statistical interactions and squared terms in the context of the binary regression model – specifically, in the context of logistic regression. As in the previous chapter, the principles discussed in this chapter apply to the other models considered in this book. That is, assessing statistical interactions, squared terms, and model diagnostics are similar in form across the general linear model for limited dependent variables. This chapter illustrates these principles across model types, and we refer readers back to this chapter throughout the remainder of the chapters focused on specific model types.

As described in the previous chapter, regression models for limited dependent variables, including logistic regression, are quite different in their interpretation than ordinary least squares regression models. The estimated coefficients in the case of a logistic regression model tells us how much change there is in the log-odds of the outcome for each unit change in each independent variable. The fact that the coefficient tells us about something other than the outcome itself implies a difficulty in interpreting interaction and squared terms. Interaction and squared terms assess whether the effect of one variable varies with the levels of another (mathematically linked as in the case of squared terms, or not). When product terms are included in a generalized linear model, the coefficient associated with the product term does not tell you whether there is significant moderation in the data (Ai and Norton 2003; Mize 2019). This is because the nonlinearity observed in the interaction or squared effect could be driven by the nonlinearities included in the link function. For this reason, modern methods for assessing statistical interactions and squared terms in the context of the GLM entails significance tests on the response metric: that is, assessing moderation in the GLM entails assessing differences in the fitted values from the model (e.g., predicted probabilities) rather than assessing the estimated coefficients.

Below, we describe how to test and probe statistical interactions in the GLM. We describe examples using the same data as the previous chapter, modeling whether the respondent feels safe at night. In this context, we describe how to assess three kinds of interaction effects: interactions between two categorical variables, interactions between a categorical variable and a continuous variable, and interactions between two continuous variables. We then transition to testing for squared terms and their interactions with other variables.

DOI: 10.1201/9781003029847-6

Moderation

Categorical × Categorical

In the previous chapter, we used logistic regression to model the log-odds of whether the respondent feels safe walking alone at night. Our interpretation focused on the effect of female on feeling safe, and we found a negative association, meaning that being female is associated with a decrease in the log-odds of reporting feeling safe at night. Now suppose we wanted to test whether the effect of gender on feeling safe at night is moderated by race; that is, whether being racially minoritized alters the effect of gender on feeling safe at night. Due to institutionalized racism, for example, an argument could be made that minoritized men might not feel as safe as majority men. Figure 6.1 shows the proportion of respondents who feel safe walking alone at night by gender and race. As illustrated, the first two columns show the disparity in feeling safe at night by gender for nonminoritized individuals. The observed difference is approximately .2. The second two columns show the disparity in feeling safe at night by gender for minoritized individuals. Here, the disparity is approximately .05. Below, we formally assess whether this difference is statistically significant.

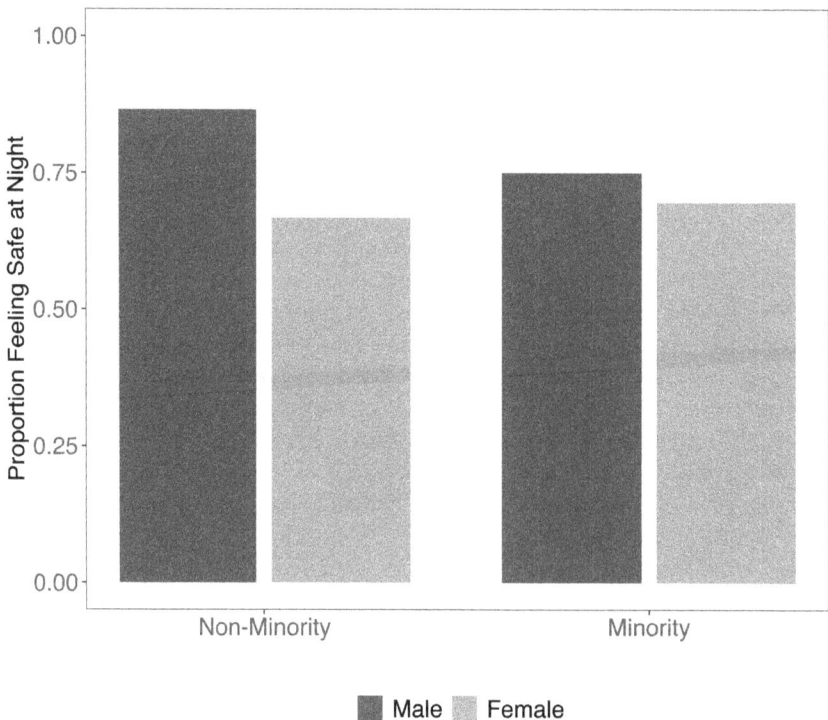

FIGURE 6.1
Proportion of respondents reporting feeling safe walking alone at night by race and sex.

Table 6.1 presents three logistic regression models predicting whether the respondent feels safe walking alone at night. Model 1 includes an interaction term between female and minority. For clarity, this is the full model specification:

$$\eta\left(Safe\right) = \alpha + \beta_{\text{religious}} \times \text{religious} + \beta_{\text{minority}} \times \text{minority} + \beta_{\text{female}} \times \text{female}$$
$$+ \beta_{\text{age}} \times \text{age} + \beta_{\text{minority} \times \text{female}} \times \left(\text{minority} \times \text{female}\right)$$

Model 1 in Table 6.1 shows that compared to racial minority members, racial majority members have a lower log-odds of reporting feeling safe at night ($\beta = -.77$, $p = .008$). Similarly, women have a lower log-odds compared to men ($\beta = -1.165$, $p < .001$). Furthermore, the coefficient for the interaction term is statistically significant, indicating an increase in log-odds for respondents that are both female and minoritized ($\beta = .901$, $p = .015$). However, we noted above that the significance of the interaction term itself cannot be used to establish moderation (Mize 2019). Instead, we need to demonstrate a significant effect on the response metric – predicted probabilities in this case – to show that the effect captured in the interaction effect is driven by differences in the predicted probabilities, rather than (possibly) due to nonlinearities in the link function.

TABLE 6.1

Summary of Three Logistic Regression Models

	Model 1	Model 2	Model 3
Religious (R)	−.020	−.035	−.062
	(.018)	(.018)	(.043)
Minority (M)	−.765**	−.255	−.209
	(.287)	(.201)	(.201)
Female (F)	−1.165***	−1.534***	−1.083***
	(.118)	(.277)	(.115)
Age	−.006*	−.004	−.004
	(.003)	(.003)	(.003)
Immigration is Good for the Economy (I)		.139***	.257***
		(.038)	(.034)
F×M	.901*		
	(.370)		
F×I		.084	
		(.046)	
R×I			−.017*
			(.007)
Intercept	2.258***	1.364***	.735**
	(.189)	(.282)	(.257)

Note: N=2,163. *p<.05, **p<.01, ***p<.001.
The outcome is whether the respondent feels safe walking alone at night (=1).

The first thing we do to assess the interaction effect is compute the predicted probabilities for each cell of the interaction effect. We do so using `margins.des` and `margins.dat`, and provide a plot to illustrate the patterns (net of controls). Below is the code, and corresponding output and graph:

```
> m2 <- glm(safe ~ religious + minority*female +
age,data=X,family="binomial")

> design <- margins.des(m2,ivs=expand.grid(minority=c(0,1),fem
ale=c(0,1)))

> pdat <- margins.dat(m2,design)

> pdat
  minority female religious     age fitted    se     ll    ul
1        0      0    3.602 53.146  0.866 0.011 0.844 0.889
2        1      0    3.602 53.146  0.751 0.050 0.653 0.849
3        0      1    3.602 53.146  0.669 0.014 0.641 0.697
4        1      1    3.602 53.146  0.699 0.050 0.601 0.796

> pdat <- mutate(pdat,Minority=rep(c("No","Yes"),2),
+                     Female=rep(c("Male","Female"),each=2),
+                     xaxs=c(-.05,0,.1,.15))
ggplot(pdat,aes(x=xaxs,y=fitted,ymin=ll,ymax=ul,group=Minority,color=M
inority)) +
  geom_pointrange() + theme_bw() + labs(x="",y="Pr(Safe at night)") +
  scale_x_continuous(breaks=c(-.025,.125),labels=c("Male","Female"),
limits=c(-.1,.2)) +
  theme(legend.position="bottom") +
  scale_color_manual(values=c("grey0","grey60"))
```

Figure 6.2 illustrates that the effect of gender is washed out for minoritized individuals. Women, regardless of minority status, are less likely to report feeling safe at night. For men, minoritized men are less likely to report feeling safe. That is, the effect of race matters among men. As noted in the previous chapter, we can use `first.diff.fitted` to compute the difference in pairs of predicted probabilities. Here, we note that we can compute multiple comparisons by giving the function multiple pairs of rows to compare, as illustrated below. Doing so, we see that the marginal effect of gender for majority members is significant (estimate = .197, $p < .001$), and that the marginal effect of gender for minoritized individuals is not significant (estimate = .052, $p = .76$). This still does not tell us whether the marginal effect of gender varies by race. For that, we need to test whether the first differences for majority members are significantly different from the first differences for minority members. To do so, we turn to the `second.diff.fitted` function, which explicitly tests the difference between two different first differences. Here is the code and results to explicitly test the interaction between gender and minority status:

```
> first.diff.fitted(m2,design,compare=c(3,1,4,2))

  fitted1 fitted2 first.diff std.error statistic p.value     ll      ul
1   0.669   0.866     -0.197     0.018   -10.752    0.00 -0.233 -0.161
2   0.699   0.751     -0.052     0.069    -0.756    0.45 -0.188  0.084

second.diff.fitted(m2,design,compare=c(3,1,4,2))

  term    est std.error statistic p.value     ll     ul
1   b1 -0.145     0.072    -2.019   0.043 -0.285 -0.004
```

The function `second.diff.fitted` takes the model object, the design matrix that generates the predicted probabilities that you are comparing, and the four rows to compare. Importantly, the first two numbers in compare are used to generate the first first difference, the last two numbers in compare are used to generate the second first difference, and the second difference is computed as the first first difference, minus the second first difference. Put differently, `second.diff.fitted`, as specified above, computes the following second difference: {Pr(design[row 1]) − Pr(design[row 3])} − {Pr(design[row2]) − Pr(design[row4])}, where Pr(design[row x]) refers to the probability that the outcome=1, conditional on the covariate values in row x of the design matrix. In this example, rows 1 and 3 refer to majority men and majority women, respectively, and rows 2 and 4 refer to minority men

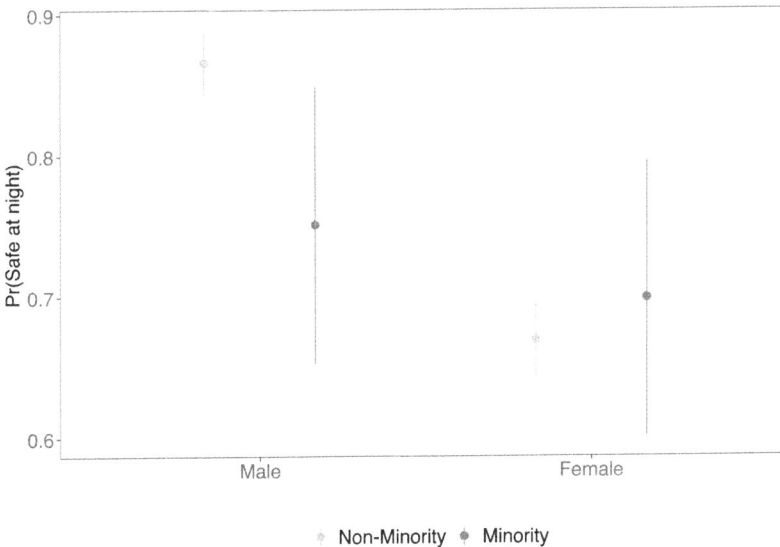

FIGURE 6.2
Marginal probability of reporting feeling safe at night. Margins drawn from Table 6.1, Model 1. 95% confidence intervals shown (via the Delta method).

and minority women, respectively. The first comparison (1,3) is the effect of female for majority individuals, and the second comparison (2,4) is the effect of female for minority individuals. As shown above, the effect of female for majority members is −.197 and the effect of female for minority members is −.052. The difference between them (i.e., the second difference) is −.145 (−.197−−.052=−.145, $p=.043$), meaning that the change in probabilities for women is weaker for minoritized individuals.

As is the case with `first.diff.fitted`, the `second.diff.fitted` function uses the Delta method to compute standard errors via the `marginaleffects` package. `Catregs` also includes the more general `compare.margins` function that includes inference or uncertainty estimated via simulation. This function is "more general" in that it works with both marginal effects at means (MEMs) and average marginal effects (AMEs). Specifically, the `compare.margins` function in `catregs` uses a simulation procedure to generate p-values associated with differences in (conditional) marginal effects (King, Tomz, and Wittenberg 2000). Given two marginal effects (conditional or not) and their standard errors, `compare.margins` simulates draws from both marginal effect distributions and compares them. The reported p-value is the proportion of times the two distributions overlapped. Below we apply this function to assess our interaction between gender and race in Model 1. We find the same result as we did with `second.diff.fitted` – the difference is still −.145, and the p-value is significant.

```
>compare.margins(margins=c(mem1$first.diff,mem2$first.
diff),margins.ses=c(mem1$std.error,mem2$std.error))

$difference
[1] -0.145

$p.value
[1] 0.018
```

When generating predicted probabilities, as reviewed in the previous chapter and above, we choose values to use for our covariates. If we set them to their mean values, we are estimating MEMs or some variation of this (e.g., setting them to theoretically meaningful values). Alternatively, we can compute the AMEs, which refers to the marginal effect on the outcome for a unit change in a focal independent variable for each observation in the data.[1] We can also use AMEs to evaluate statistical interactions. Doing so entails computing separate conditional AMEs (conditional on one of the variables in the interaction), and then comparing them just like we did with the marginal effects at means.

Conditional AMEs are the marginal effect on the outcome for a focal independent variable within subsets of the data defined by the conditioning variable. For example, the AME of female conditional on being minoritized can be defined as: $\Pr(y=1\,|\,\textit{female}=0 \textit{ and minority}=1, x_k=x_{ki})-\Pr(y=1\,|\,\textit{female}=1 \textit{ and}$

minority $= 1$, $x_k = x_{ki}$) and the AME of female conditional on being a majority member can be defined as: $\Pr(y = 1 \mid female = 0 \text{ and } minority = 0, x_k = x_{ki}) -$ $\Pr(y = 1 \mid female = 1 \text{ and } minority = 0, x_k = x_{ki})$. Note that in both conditional AMEs, the value of minority does not change. These can be computed using the marginaleffects package and the following R call:

```
> summary(marginaleffects(m2,variables="female",newdata=datagr
id(minority=0)))
    Term  Effect Std. Error z value   Pr(>|z|)   2.5 % 97.5 %
1 female -0.2032    0.01952  -10.41 < 2.22e-16 -0.2415 -0.165
```

```
> summary(marginaleffects(m2,variables="female",newdata=datagr
id(minority=1)))
    Term   Effect Std. Error z value Pr(>|z|)   2.5 % 97.5 %
1 female -0.05273    0.07022 -0.7509  0.45269 -0.1904 0.0849
```

The AME of female for majority members is −.2032, meaning that the probability a female majority member reports feeling safe at night is .2 less than the probability a male majority member feels safe at night. To the second decimal place this is the same value as we reported above for the marginal effect of female at means (i.e., −.197). The AME of female for minority members is −.0527, meaning that the probability a female minority member reports feeling safe at night is .05 less than the probability a male minority member feels safe at night. To the second decimal place, this is the same value as we reported above for the marginal effect of female at means (i.e., −.052).

The marginaleffects package computes the conditional AMEs by implementing the newdata option. We can compare the conditional AMEs to formally assess the interaction using compare.margins. We do so below, specifying that the margins to compare are the conditional AMEs and their estimates of uncertainty.

```
> ma1 <- summary(marginaleffects(m2,variables="female",newdata
=datagrid(minority=0)))
```

```
> ma2 <- summary(marginaleffects(m2,variables="female",newdata
=datagrid(minority=1)))
```

```
> cames <- rbind(ma1,ma2)
```

```
> cames
    Term   Effect Std. Error  z value Pr(>|z|)   2.5 %   97.5 %
1 female -0.20321    0.01952 -10.4097  < 2e-16 -0.2415 -0.1650
2 female -0.05273    0.07022  -0.7509  0.45269 -0.1904  0.0849

Model type:  glm
Prediction type:  response
```

```
> compare.margins(margins=cames$estimate,margins.
ses=cames$std.error)
$difference
[1] -0.15

$p.value
[1] 0.017
```

Above, we first define the two conditional AMEs as objects (ma1 and ma2). We then put them together into a matrix called cames. compare.margins subtracts the second conditional AME from the first (i.e., row 1−row 2).[2] On average, the effect of female on the probability of feeling safe at night is .15 less for minoritized individuals than it is for majority members ($p=.017$). Again we note the similarities in point estimates: the second difference in MEMs is −.145 ($p=.043$) and the difference in conditional AMEs is −.15 ($p=.017$).

Categorical × Continuous

Our second example of moderation considers the effect of whether the respondent thinks immigration is good for the economy on whether they feel safe walking alone at night. We might reason that fear of the unknown drives both, so thinking immigrants are bad for the economy might be related to being fearful at night. We assess whether this relationship is moderated by gender. Model 2 in Table 6.1 presents the logistic regression parameter estimates (in log-odds form). We note that the interaction effect in Model 2 does not reach the conventional level of statistical significance ($p=.069$).

Figure 6.3 illustrates the predicted probabilities for males and females as the immigration variable varies from its minimum (0) to its maximum (10). Given the design matrix used to generate Figure 6.3, we can compute pairs of first differences by gender using first.diff.fitted. Likewise, we can compare the effect of gender at any level of immigration is good for the economy using second.diff.fitted. Below we show the code used to generate the design matrix and then print it. We evaluate the change in predicted probabilities by gender conditional on the immigration variable being set to 0. We find that the predicted probability is .36 lower for females when immigration is good for the economy is set to its minimum (comparing rows 2 and 1). Next, we find that the predicted probability is .07 lower for females when immigration is good for the economy is set to its maximum (comparing rows 22 and 21). The first test of second differences shows that the difference in predicted probabilities by gender is significantly different when the immigration variable is set to its minimum versus its maximum (−.291, $p<.001$). The second test of second differences shows that the difference in predicted probabilities by gender is significantly different when the immigration variable is set to 4 versus 9 (−.146, $p<.001$).

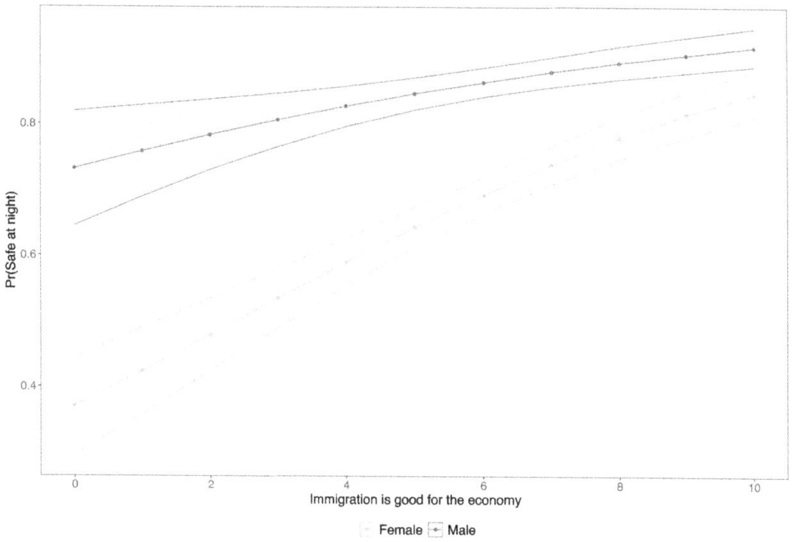

FIGURE 6.3
Marginal probability of reporting feeling safe at night. Margins drawn from Table 6.1, Model 2. 95% confidence intervals shown (via the Delta method).

```
> design <- margins.des(m2,ivs=expand.
grid(female=c(0,1),immigration.good.economy=0:10))

> round(design,3)
   female immigration.good.economy religious minority     age
1       0                        0     3.601    0.076 53.105
2       1                        0     3.601    0.076 53.105
3       0                        1     3.601    0.076 53.105
4       1                        1     3.601    0.076 53.105
5       0                        2     3.601    0.076 53.105
6       1                        2     3.601    0.076 53.105
7       0                        3     3.601    0.076 53.105
8       1                        3     3.601    0.076 53.105
9       0                        4     3.601    0.076 53.105
10      1                        4     3.601    0.076 53.105
11      0                        5     3.601    0.076 53.105
12      1                        5     3.601    0.076 53.105
13      0                        6     3.601    0.076 53.105
14      1                        6     3.601    0.076 53.105
15      0                        7     3.601    0.076 53.105
16      1                        7     3.601    0.076 53.105
17      0                        8     3.601    0.076 53.105
18      1                        8     3.601    0.076 53.105
19      0                        9     3.601    0.076 53.105
20      1                        9     3.601    0.076 53.105
21      0                       10     3.601    0.076 53.105
22      1                       10     3.601    0.076 53.105
```

```
> first.diff.fitted(m2,design,compare=c(2,1,22,21))

   first.diff std.error statistic p.value      ll      ul
1      -0.361     0.058    -6.220   0.000  -0.475  -0.247
2      -0.070     0.023    -3.003   0.003  -0.116  -0.024

> second.diff.fitted(m2,design,compare=c(2,1,22,21))

   term     est std.error statistic p.value      ll      ul
1    b1  -0.291     0.074    -3.905       0  -0.437  -0.145

> second.diff.fitted(m2,design,compare=c(10,9,20,19))

   term     est std.error statistic p.value      ll      ul
1    b1  -0.146     0.032    -4.595       0  -0.208  -0.084
```

We can aggregate over multiple runs of `second.diff.fitted` to assess whether the effect of attitudes toward immigration varies by gender. We find that, on average, for each unit increase in positive attitudes toward immigration the probability of women reporting feeling safer at night is .047 (s.e. = .006, $p < .001$), and the probability of men reporting feeling safer at night is .019 (s.e. = .007, $p = .011$); the average second difference .029 is significant (i.e., .047–.019; s.e. = .009, $p = .002$), meaning that the return to favorable attitudes toward immigrants on feeling safe at night is significantly stronger for women.

We do not show the code for the aggregating first and second differences here. It is available online. But the logic is straightforward. We compute first and second differences for each adjacent category of the immigration variable. The estimate is the mean of the pooled differences. To compute the standard error of the average estimate, we use Rubin's rule to combine standard errors using the `rubins.rule` function in `catregs`. We describe this function in more detail in Chapter 11.

Next, we show how to illustrate the interaction between attitudes toward immigration and gender using conditional AMEs. This is relatively straightforward to implement. We compute the conditional AME of the immigration variable, conditioning on respondent gender. We then compare them using `compare.margins`, as follows:

```
> ma1 <- summary(marginaleffects(m2,variables="immigration.
good.economy",newdata=datagrid(female=0)))

> ma2 <- summary(marginaleffects(m2,variables="immigration.
good.economy",newdata=datagrid(female=1)))

> cames <- rbind(ma2,ma1)

> cames
```

```
                        Term  Effect Std. Error z value   Pr(>|z|)
2.5 %  97.5 %
1 immigration.good.economy 0.04798   0.005717   8.392 < 2.22e-16
0.036772 0.05918
2 immigration.good.economy 0.01662   0.004448   3.737 0.00018622
0.007903 0.02534

Model type:  glm
Prediction type:  response

> compare.margins(margins=cames$estimate,margins.
ses=cames$std.error)

$difference
[1] 0.031

$p.value
[1] 0
```

We find that the conditional AME of attitudes toward immigration for males is .017 (the MEM was .019) and for females it is .048 (the MEM was .047). The difference between these conditional marginal effects tells us how the effect of the immigration variable varies by gender. We find that the effect is stronger for females (.031, $p < .001$; the MEM was .029 above). Note that the log-odds coefficient for the interaction term is not statistically significant, but the MEMs and the conditional AMEs are both significant.

Continuous × Continuous

We next describe how to illustrate a statistical interaction between continuous variables. Model 3 in Table 6.1 summarizes a model with an interaction between religiosity and the immigration variable from the previous section (increasing values mean the respondent thinks immigration is good for the economy). Figure 6.4 illustrates how the predicted probabilities of feeling safe at night vary with our two independent variables. As the immigration variable increases, respondents feel safer at night. However, this effect gets weaker as religiousness increases.

The plot above illustrates the patterns associated with the interaction effect, but it does not formally test the interaction on the metric of the variable itself. Doing so with MEMs entails computing the first differences for a focal independent variable at several levels of the other independent variable and then aggregating over the focal variable's estimates of first differences at the other variable. Once that is completed, we can compare the aggregated MEMs. Below we provide code for generating and aggregating the MEMs. We begin by defining the design matrix. We compute the first difference for each adjacent category of the immigration variable at each level of religion. fd1 computes the first differences when religion=0. We then create a loop

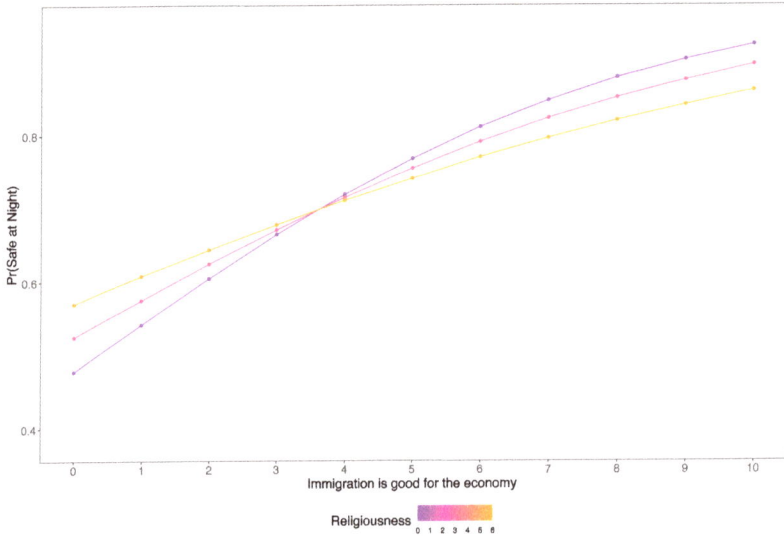

FIGURE 6.4
Marginal probability of reporting feeling safe at night. Margins drawn from Table 6.1, Model 3. 95% confidence intervals shown (via the Delta method).

to vary values of religion. Each time R passes through the loop, it adds 1 to the value of religion, recomputes fd1, and then appends the new estimates to the old ones. After the loop, we aggregate the first differences by taking the mean of the estimates and applying Rubin's rule to the standard errors.

```
> design <- margins.des(m2,ivs=expand.grid(immigration.good.
economy=0:10,religious=0))

> fd1<-first.diff.fitted(m2,design,compare=
c(11,10,10,9,9,8,8,7,7,
6,6,5,5,4,4,3,3,2,2,1))

> fd1<-mutate(fd1,relig=0)

> for(i in 1:10){
+   design <- margins.des(m2,ivs=expand.grid(immigration.good.
economy=0:10,religious=i))
+   fd2<-first.diff.fitted(m2,design,compare=
c(11,10,10,9,9,8,8,7,7,6,
6,5,5,4,4,3,3,2,2,1))
+   fd2<-mutate(fd2,relig=i)
+   fd1<-rbind(fd1,fd2)}

> fds<- fd1 %>% group_by(relig) %>%
summarize(fds=mean(first.diff),ses=rubins.rule(std.error))

> fds
```

```
# A tibble: 11 × 3
   relig      fds      ses
   <dbl>    <dbl>    <dbl>
1      0  0.0444  0.00726
2      1  0.0422  0.00638
3      2  0.0397  0.00549
4      3  0.0372  0.00481
5      4  0.0347  0.00482
6      5  0.0318  0.00541
7      6  0.029   0.00601
8      7  0.0262  0.00707
9      8  0.023   0.00798
10     9  0.0202  0.00922
11    10  0.017   0.0105
```

Figure 6.5 shows the aggregated MEMs for the immigration variable as religiousness varies; that is, Figure 6.5 is a plot of fds. We can see in the figure that the effect of immigration is stronger when religiousness is low. More formally, comparing the aggregated MEMs, we find that the effect of religion is stronger when immigration is 0 than when it is 10 (.044–.017 = .027, $p = .014$). This comes from the following code:

```
> compare.margins(margins=fds$fds[c(1,11)],margins.ses=fds$ses[c(1,11)])
$difference
[1] 0.027

$p.value
[1] 0.014
```

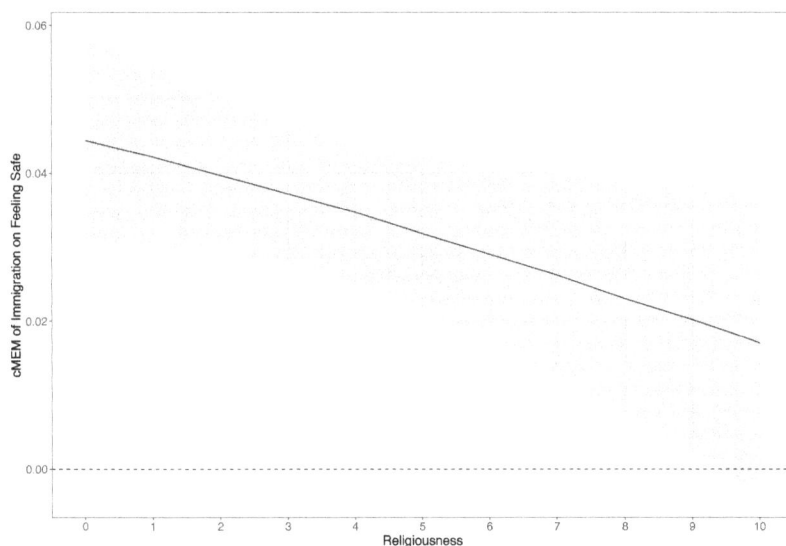

FIGURE 6.5
Conditional marginal effect at means of immigration on feeling safe walking alone at night.

Of course we can do the same things with AMEs. Below is the code to generate the conditional AMEs. We again use a loop to streamline the code by varying levels of religiousness.

```
> ma1 <- summary(marginaleffects(m2,variables="immigration.good.
economy",newdata=datagrid(religious=0)))

> for(i in 1:10){
+    ma2 <- summary(marginaleffects(m2,variables="immigration.
good.economy",newdata=datagrid(religious=i)))
+    ma1<-rbind(ma1,ma2)}

> ma1 <- mutate(ma1,relig=0:10)

> round(ma1[,3:ncol(ma1)],3)
```

```
    estimate std.error statistic p.value conf.low conf.high relig
1      0.040     0.005     7.905   0.000    0.030     0.050     0
2      0.038     0.004     8.731   0.000    0.030     0.047     1
3      0.036     0.004     9.394   0.000    0.029     0.044     2
4      0.034     0.004     9.464   0.000    0.027     0.041     3
5      0.032     0.004     8.614   0.000    0.025     0.040     4
6      0.030     0.004     7.123   0.000    0.022     0.038     5
7      0.027     0.005     5.558   0.000    0.018     0.037     6
8      0.025     0.006     4.227   0.000    0.013     0.036     7
9      0.022     0.007     3.177   0.001    0.009     0.036     8
10     0.019     0.008     2.362   0.018    0.003     0.036     9
11     0.016     0.010     1.727   0.084   -0.002     0.035    10
```

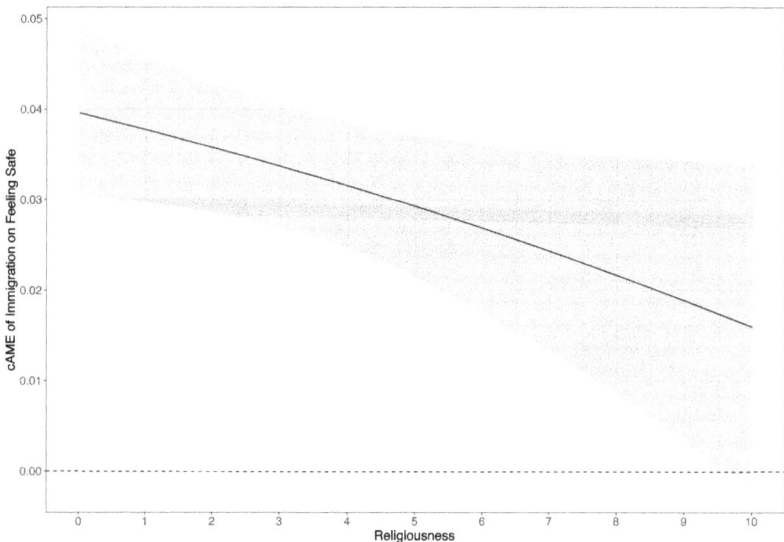

FIGURE 6.6
Conditional average marginal effect of immigration on feeling safe walking alone at night.

Figure 6.6 is a plot of ma1. It shows the same pattern as Figure 6.5: the effect of the immigration variable is stronger at lower levels of religiousness. Testing the interaction effect entails demonstrating that the conditional AME of immigration significantly varies by levels of religiousness. As shown below, the cAME of immigration when religiousness is 0 (.04) is indeed larger than the cAME of immigration when religiousness is 10 (.016; .04−.016=.024, p=.01).

```
> compare.margins(margins=ma1$estimate[c(1,11)],margins.
ses=ma1$std.error[c(1,11)])
$difference
[1] 0.024

$p.value
[1] 0.013
```

The statistics described above evaluate the interaction term in the model. A heatmap or contour plot (Mize 2019) is a useful way to illustrate the substantive implications of the interaction term. The x and y axes in the graph are defined by the two continuous variables and the plot illustrates how the predicted probabilities change in the two-dimensional space that is defined by the two variables in the interaction. Figure 6.7 is such a heat map illustrating the interaction between religiousness and the extent to which the respondent thinks immigration is good for the economy. It shows that as religiousness increases, the effect of attitudes toward immigration decreases. Toward the bottom of the plot, the predicted probabilities increase more than they do toward the top of the plot (where they start higher and end lower).

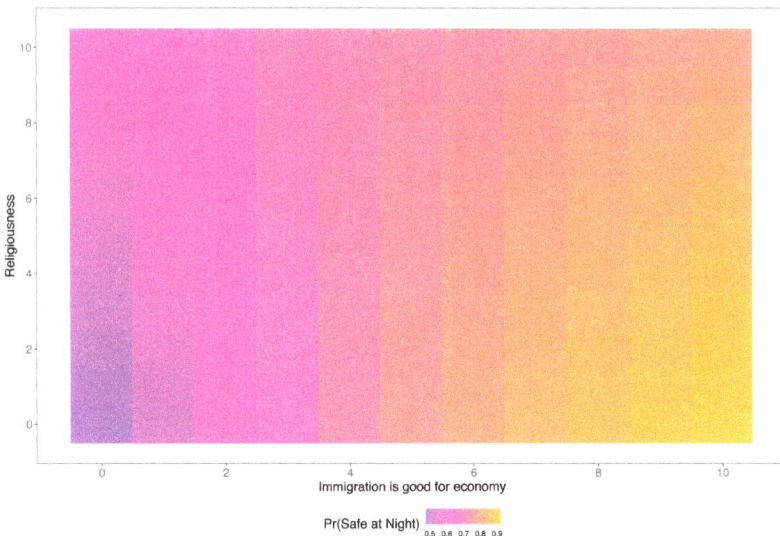

FIGURE 6.7
Heatmap illustrating the probability of feeling safe walking alone at night.

Squared Terms

In this section, we describe how to assess polynomial terms in the context of regression models for limited dependent variables. We use the same data source as our previous examples (respondents from the United Kingdom's European Social Survey), but model whether the respondent has a high income, defined as being in the top 30% of the income distribution. We first illustrate a model that includes a squared term for age. We then include a squared term in an interaction effect with whether the respondent is married to illustrate how to interpret such a model. Table 6.2 presents the log-odds coefficients for two logistic regression models. Model 1 shows that the log-odds coefficient for age-squared is significant, and Figure 6.8 shows the predicted probabilities as age varies by whether the model includes age as a linear term (full model not shown) or as a squared term (as in Table 6.2, Model 1). In the code for Model 1 below, note how to specify a polynomial term in R via the I function and the corresponding exponent:

```
m2 <- glm(highinc ~ religious + minority  + female  + married
+ age + I(age^2) ,data=X2,family="binomial")
```

TABLE 6.2

Summary of Two Logistic Regression Models

	Model 1	Model 2
Religious	−.011	−.010
	(.018)	(.018)
Minority	−.037	−.054
	(.196)	(.194)
Female	−.425***	−.441***
	(.103)	(.103)
Married=1 (M)	−.775***	.469
	(.226)	(2.911)
Age (A)	.170***	.179***
	(.020)	(.021)
Age-Squared (S)	−.002***	−.002***
	(.000)	(.000)
M×A		−.096
		(.106)
M×S		.001
		(.001)
Intercept	−4.241***	−3.686***
	(.521)	(.504)

Note: N=2,163. *p <.05, **p <.01, ***p <.001.
The outcome is whether the respondent has a high income (=1).

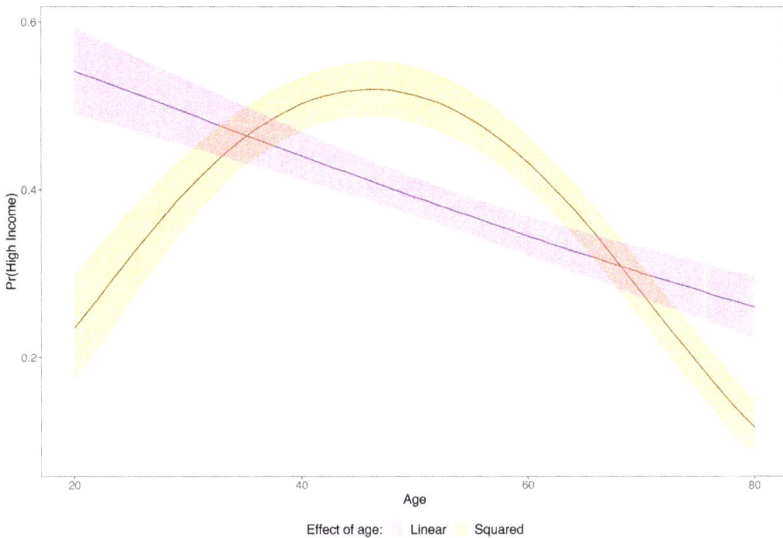

FIGURE 6.8
Predicted probabilities of having a high wage as age varies. Two different model specifications shown.

As with interaction terms, the significance of the log-odds parameter estimate for the squared term does not (necessarily) imply that the squared term has a significant impact on the predicted probabilities. While Figure 6.8 makes the need for a curvilinear relationship clear, it may be less clear in other applications. We can explicitly test whether the marginal effect of age varies over intervals of age. For example, we find that the average MEM of age from 20 to 45 is positive (MEM=.011, s.e.=.002, $p<.001$) and we find that the average MEM of age from 50 to 80 is negative (MEM=−.013, s.e.=.003, $p=.4$). Applying the compare.margins function to the average MEMs, we find that the difference is statistically significant ($p<.001$).

The effect of age in terms of changes to the log-odds in Model 1 of Table 6.2 is captured by two different parameter estimates. We can use MEMs and AMEs to identify the effect of a one unit change in age on the probability of reporting a high income. The MEMs and the AMEs adjust for the fact that age is squared by shifting both the linear and the squared effect of age simultaneously (as they are mathematically linked). We find that, on average, with covariates set to their means, a one unit change in age is associated with a −.006 decrease in the probability of having a high income (MEM=−.006, s.e.=.002, $p=.02$). Similarly, we find that, on average, in the observed data, a one unit change in age is associated with a −.003 decrease in the probability of having a high income (AME=−.003, s.e.=.001, $p<.001$). We note that the MEM is sensitive to the values used to estimate the first differences, while the AME is not since it uses the observed data. This is due to the fact that the analyst sets the values to use in the design matrix when working with MEMs.

Model 2 in Table 6.2 includes an interaction between married and age. We find that the main effect of age does not vary by married (β=−.096, s.e.=.106, p=.37), and neither does the squared term (β=.001, s.e.=.001, p=.17). However, the statistical significance of an interaction term does not tell whether moderation is observed on the response metric. As illustrated in Figure 6.9, it seems middle-aged people who are not married are significantly more likely to have a high wage than middle-aged married people, but this difference disappears at younger and older ages. We evaluate this more formally using MEMs and AMEs below.

Figure 6.10 shows how the effect of marriage changes with age. Panel A illustrates the marginal effect of marriage with covariates set to their mean and Panel B illustrates the average marginal effect of marriage with covariates set to their observed values. We use compare.margins to test whether the MEM for marriage conditional on age=40 is different from the MEM for marriage conditional on age=65, and we find that the effect is more negative at 40 (est=−.217, p=.008).[3] Similarly, we use compare.margins to test whether the AME for marriage conditional on age=40 is different from the AME for marriage conditional on age=65, and we again find that the effect is more negative at 40 (est=−.247, p=.009). Either of these two last statements is sufficient evidence to demonstrate the interaction between age-squared and marriage, provided we have already demonstrated the need for a squared term.

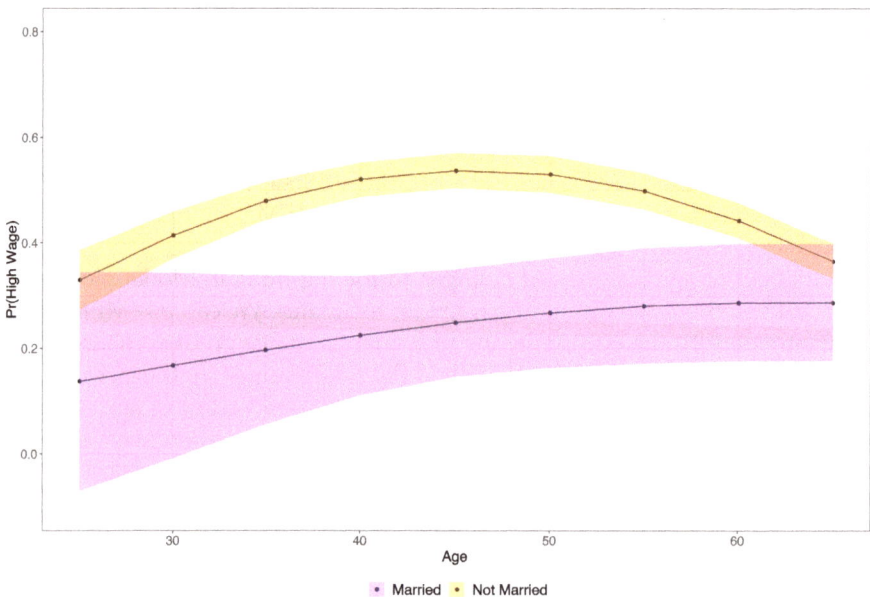

FIGURE 6.9
Marginal probability of having a high income. Margins drawn from Table 6.2, Model 2. 95% confidence intervals shown (via the Delta method).

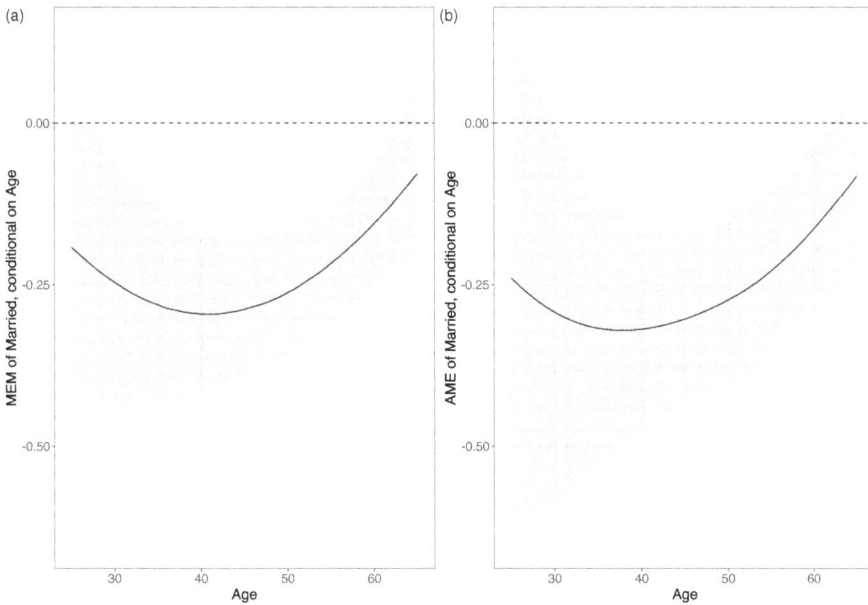

FIGURE 6.10
(a) Conditional marginal effect at means of being married on feeling safe walking alone at night and (b) conditional average marginal effect of immigration on feeling safe walking alone at night.

Notes

1. Here and throughout, we adopt a generic description of margins as representing a "unit change." For categorical variables in particular, this refers to the average discrete change (Long and Mustillo 2021). We discuss issues of scaling in more detail in Chapter 11.
2. Of course you can give compare.margins a matrix and select any two rows for comparison using base R commands.
3. The values of 40 and 65 were selected by visual inspection of Figure 6.9 and the underlying data.

7

Regression for Ordinal Outcomes

Ordinal outcomes are those where response categories are ordered in some substantive way but the distances between categories are undefined. How much better is "excellent" than "very good" on an evaluative measure, for example, and is it the same amount of better as "very bad" to "terrible?" These types of measures are ubiquitous in social scientific research, from commonly used Likert-type scales to more well-defined yet still ordinal categories like levels of education (e.g., "High School or Less," "Some college," "College degree," and "Graduate degree"). In this chapter, we address models designed for these types of outcomes, show their relationship to the binary regression models discussed in the previous chapters, and discuss their relationship to models that do not assume inherent ordering in response categories – that is, multinomial logistic regression models.

Generalizing the BRM

As with the BRM, the ordinal regression model (ORM) can be derived statistically or theoretically, each leading to the same model. The statistical approach generalizes the BRM's statistical approach to examine the probability of a response being less than or equal to category m versus being greater than m given $\mathbf{x}$. Indeed, the BRM is a special case of the ORM where there are only two categories. Starting with a preferred link function, logit or probit, we generalize:

$$\eta(y) = \mathbf{Xb} \tag{7.1}$$

where η is the link function and $\mathbf{Xb}$ is the matrix of covariates and their coefficients, to:

$$\eta_{>m|\leq m}(y) = \eta(\tau_m + \mathbf{Xb}) \tag{7.2}$$

for $m=1$ to $J-1$ where J is the highest category and τ_m is the category-specific intercept for category m. With some rearranging to predict the probability of a response being less than or equal to m versus greater than m rather than the

DOI: 10.1201/9781003029847-7

reverse (because it does not make sense to predict a response being greater than m for the last category since the predicted probabilities for all categories of m have to sum to 1), the model is more typically presented as:

$$\eta_{\leq m|>m}(y) = \eta(\tau_m - \mathbf{X}b) \tag{7.3}$$

Maximum likelihood estimation then jointly maximizes the log-likelihood of the sum of all categories of m. Given this general form, we can get the probability of any given category by subtracting the previous category:

$$\eta_m(y) = \eta(\tau_m - \mathbf{X}b) - \eta(\tau_{m-1} - \mathbf{X}b) \tag{7.4}$$

An assumption of this generalization of the BRM is that the coefficients for all categories of m are the same. This assumption is often called the *parallel regression assumption*, or the *proportional odds assumption* if the link function is the logit, which we discuss at the end of this chapter. The assumption makes more sense and is clearer under the alternative derivation of the ORM based on the latent-variable approach.

Like the BRM, the latent-variable approach to deriving the ORM conceptualizes the ordinal model as a problem of estimating multiple thresholds for various empirical manifestations of the same underlying propensity y^*. As with the BRM, each observed outcome represents whether the latent variable y^* is above or below a threshold τ_m. Therefore, the relationship between the observed outcome y and the unobserved propensity y^* is:

$$y_i = \begin{cases} 1 & \text{if} & -\infty \leq y_i^* < \tau_1 \\ 2 & \text{if} & \tau_1 \leq y_i^* < \tau_2 \\ 3 & \text{if} & \tau_2 \leq y_i^* < \tau_3 \\ \vdots & & \vdots \\ J & \text{if} & \tau_{J-1} \leq y_i^* < \infty \end{cases} \tag{7.5}$$

As with the BRM, the error term cannot be directly estimated because y^* is unobserved. As such, the relationship between y^*, Var(ε), and the predicted probability of a case being in a given ordinal category changes as a function of the value of y^* relative to a given threshold and the assumed variance distribution. Two commonly assumed distributions of the error term are Var(ε)=1 and Var(ε)=$\pi^2/3$, corresponding to the ordered probit and ordered logit models, respectively. Figure 7.1 is an example where there are four response categories and thus three thresholds; this is similar to the relationship shown in

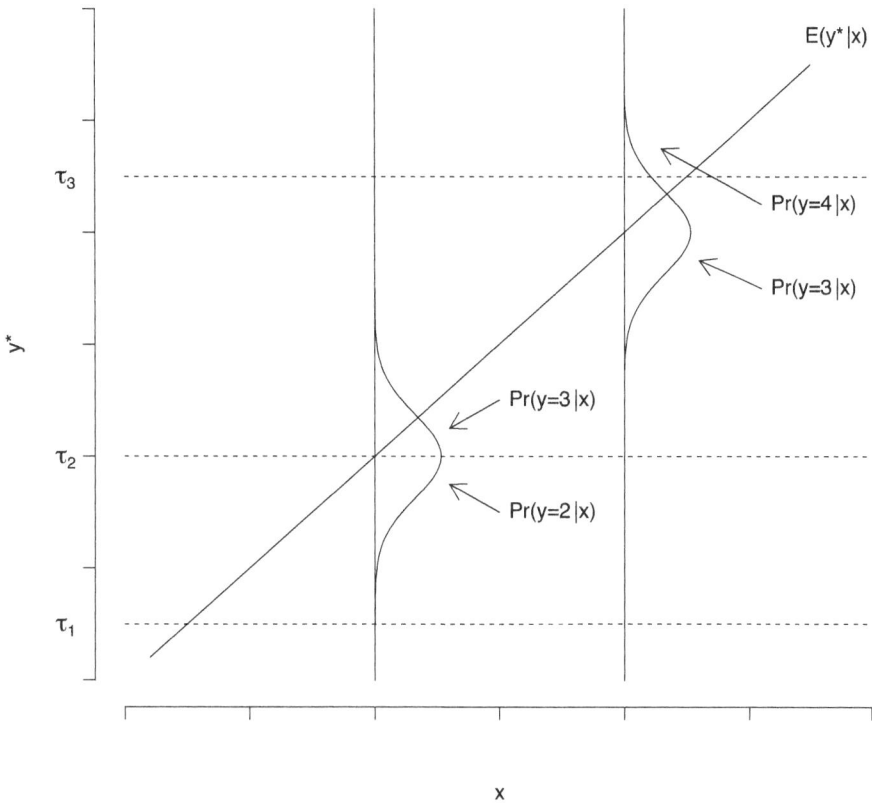

FIGURE 7.1
Relationship between y^*, Var(ε), and Pr($y = m$) in the latent-variable approach.

Figure 5.2, but with more threshold values. Indeed, the BRM can be thought of as a special case of the ORM with only two categories.

Given that the ORM is a generalization of the BRM, estimating an ORM with only two categories produces the same resulting model as estimating the BRM. Indeed, the only difference between the two models is a difference in how the intercept is handled. Recall that we arbitrarily set the threshold τ to be 0 in the BRM. In the ORM, we estimate the thresholds and instead assume the intercept of the model to be 0. Accordingly, the intercept reported in R for the BRM will be of the same magnitude but have a different sign than the threshold τ_1 reported for the same model estimated as an ORM. This is just a difference of notation and convention and all variable coefficients and resulting predicted probabilities are the same across the two models.

Estimating an ORM in R

Estimating an ORM in R requires the `polr` function from the MASS package. Using a call similar to the one used to estimate the BRM, below we reproduce the R output and then provide a formatted table of the output (Table 7.1):

```
m1<-polr(walk.alone.dark ~ religious + minority  + female +
age + emp1 + emp2,data=dat, Hess=TRUE)
summary(m1)
Call:
polr(formula = walk.alone.dark ~ religious + minority + female
+     age + emp1 + emp2, data = dat, Hess = TRUE)

Coefficients:
             Value Std. Error t value
religious -0.033715    0.014167   -2.380
minority  -0.220738    0.155629   -1.418
female    -0.982623    0.085786  -11.454
age       -0.005491    0.002352   -2.335
emp1       0.485974    0.226040    2.150
emp2       0.776218    0.245262    3.165

Intercepts:
                     Value    Std. Error t value
Very unsafe|Unsafe  -3.1695    0.2574    -12.3134
Unsafe|Safe         -1.6482    0.2478     -6.6516
Safe|Very safe       0.5716    0.2452      2.3307

Residual Deviance: 5006.31
AIC: 5024.31
```

Above, we use the original ordinal outcome that we binarized to illustrate the BRM. The dependent variable, `walk.alone.dark`, asks respondents how safe they feel walking at night, with response categories for "very unsafe," "unsafe," "safe," and "very safe." We use the same independent variables as used in Chapter 5, the respondent's religiosity, whether they are a racial/ethnic minority, whether they are female, their age in years, and whether they are traditionally employed or self-employed compared to unemployed (reference category). By default, the function assumes a `logit` link. We can estimate an ordered probit using the same call, but adding a `method="probit"` option. The full call would be:

```
m1probit <- polr(walk.alone.dark ~ religious + minority  +
female + age + emp1 + emp2,data=dat, Hess=TRUE,
method="probit")
```

TABLE 7.1

Summary of an Ordinal Logistic Regression Model
Predicting Feeling Safe Walking Alone at Night

Religious	−.034*
	(.014)
Minority (=1)	−.221
	(.156)
Female (=1)	−.983***
	(.086)
Age	−.005*
	(.002)
Traditionally Employed[1]	.486*
	(.226)
Self-Employed[1]	.776**
	(.245)
Intercepts	
Very Unsafe – Unsafe	−3.170***
	(.257)
Unsafe – Safe	−1.648***
	(.248)
Safe – Very Safe	.572*
	(.245)

Note: *p<.05, **p<.01, ***p<.001. [1]Reference category is unemployed. Source is the European Social Survey, UK Sample.

The `polr` function also allows loglog, cloglog, and cauchit links. The log–log and complementary log–log links are often used for latent variables with extreme values at the maximum and minimum categories. The cauchit link assumes that the error term follows a Cauchy distribution. We also specify `Hess=TRUE`. This saves the Hessian, which is needed for post-estimation commands that require the covariance matrix. By default, `polr` does not save this matrix.

Much of the output from the ORM is the same as the output from the BRM with a few notable exceptions. First, the model summary does not return *p*-values. These can be directly returned using built-in functions and the *t*-value of the coefficients (e.g., `dt` computes the density of a *t* distribution). Alternatively, they are provided in the output for the `catregs` function `list. coef`, which we go over in more details in the next sections. A second difference is the "Intercepts" provided under the regression coefficients. These correspond to the cut points discussed above. The `Very unsafe|Unsafe` intercept is the estimated value of the cut point between the "Unsafe" response and the "Very unsafe." Accordingly, `Unsafe|Safe` is the estimated cut point between the "Safe" and "Unsafe" responses, and `Safe|Very safe` is the estimated cut point between the "Very safe" and "Safe" responses. These cut points are typically not interpreted.

Interpretation

As with the BRM, we can proceed in interpreting the ORM using the regression coefficients, the odds ratio for the ordered logit model, and/or predicted probabilities. We walk through each below. Because the theoretical model for the ORM clearly involves a linear latent variable, more of a case can be made for interpreting y^* than with the BRM. However, the same issues with the latent propensity being unobserved apply. Accordingly, we still suggest using predicted probabilities and marginal effects, all else being equal.

Regression Coefficients

Although we are typically not substantively interested in the transformation of our ordinal dependent variable (e.g., log-odds for the ordered logit model, z-score of the probability for the ordered probit), the latent-variable interpretation of our model is more substantively interesting with the ORM than with the BRM. Recall that the BRM can be thought of as a special case of the ORM where there are only two possible values (feeling safe versus not feeling safe). In this case, the probability of feeling safe versus not provides a natural metric to interpret the model, and it is substantively interesting because it directly reflects the underlying propensity to feel safe at night. This one-to-one mapping of the latent variable to the observed outcome is less clear when we have more than two possible values (very unsafe, unsafe, safe, and very safe). For most purposes, we are less interested in the probability of feeling "Very safe" per se.[1] Rather, we are interested in what predicts feeling more versus less safe at night. Continuing our examples from Chapter 5, the coefficient for the female variable would be interpreted along these lines:

> Being female is associated with a −0.98 change in the log-odds of feeling safer at night, all else constant.

Unlike our log-odds interpretation in the BRM, we note that the log-odds reflect feeling safer, and thus selecting values of the dependent variable representing feeling safer, versus feeling less safe. We can speak to the direction and statistical significance of these coefficients, but without standardizing the coefficients in some way (Long and Freese 2006), what a −.98 change in the log-odds means is not substantively meaningful. As a first step, it may be useful enough to know which variables significantly relate to feeling safer at night. One way to quickly show this information is to use a coefficient plot. The output from list.coef can be used with ggplot to create a figure like Figure 7.2, showing the log-odds coefficients with confidence intervals.

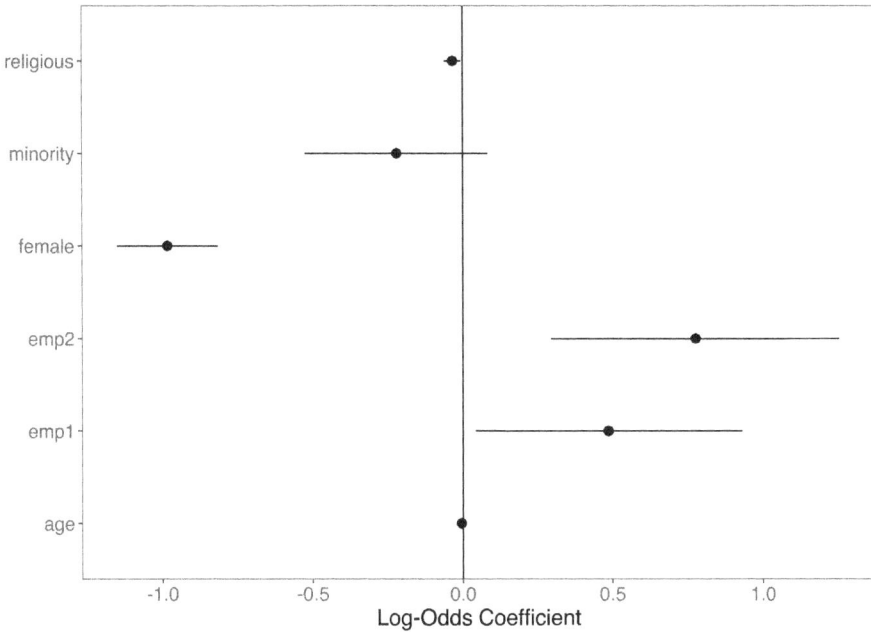

FIGURE 7.2
Coefficient plot from ORM.

Using some options to reorder the plot based on the size of the coefficients and adding variable labels, we can use the following call to create a more readily interpretable coefficient plot from our model in Figure 7.3:

```
ggplot(m1out,aes(y=reorder(variables,b),x=b,xmin=ll,xmax=ul)) +
   theme_bw() + geom_pointrange() + geom_vline(xintercept=0) +
   labs(x="Log-Odds Coefficient",y="") +
   scale_y_discrete(labels=c("female"="Female",
                             "emp1"="Traditional Employment",
                             "age"="Age",
                             "religious"="Religious",
                             "minority"="Ethnic Minority",
                             "emp2"="Self-Employed"))
```

As shown in the figure, being female has a large and significant negative effect on feeling safe at night. Being employed, both traditionally and self-employed, has a large and significant positive effect on feeling safe at night. Being a racial and ethnic minority has a negative but nonsignificant effect. It is less clear from the coefficient plot whether the effects of age and religiosity are significant, but both are negative. This is in part because both variables have larger ranges than the indicator variables with larger effects. Returning

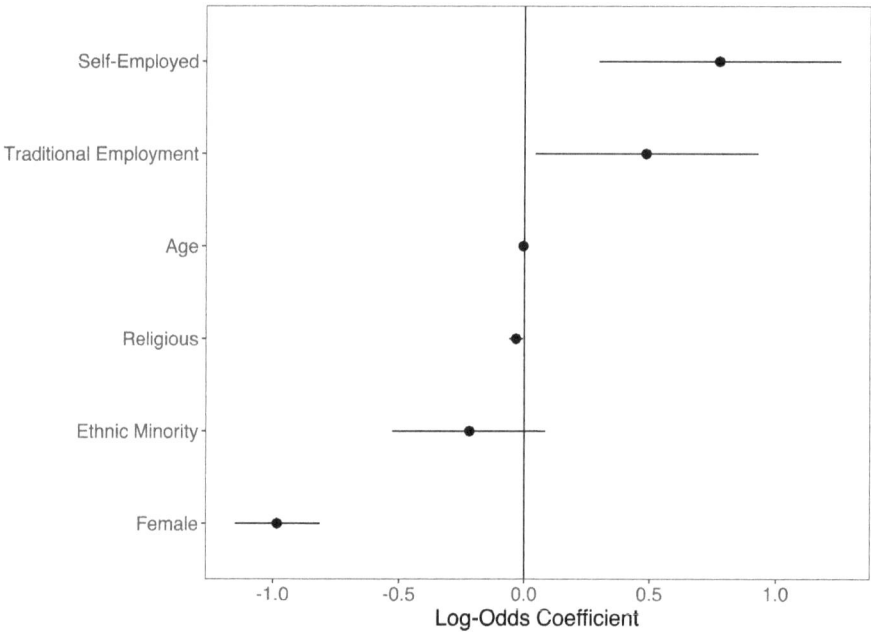

FIGURE 7.3
Reordered and labeled coefficient plot from ORM.

to our output, we see that both age and religiosity have statistically significant effects, although the magnitude of these effects (on the log-odds) is small. Before concluding that these are substantively small effects, however, we examine alternative interpretation schemes.

Odds Ratios

As with the BRM, the ordered logit model can also be interpreted using the odds ratio. Again, it is possible (but unclear why anyone would) to calculate an odds ratio for the ordered probit, but it is preferable to just use predicted probabilities for the ordered probit. To get the odds ratio, we exponentiate the ordered logit coefficients (see Chapter 5 for the math behind these calculations). Going back to the `female` variable example, we can interpret it as:

> Being female is associated with a decrease in the odds of reporting feeling safer at night by a factor of .357 ($e^{-0.983}$ =.374), all else constant.

Again, we maintain our language of interpreting the odds of selecting responses corresponding to feeling safer versus feeling less safe. We can obtain the odds ratio for all coefficients in our model using the `list.coef` function:

```
list.coef(m1)
$out
```

	variables	b	SE	z	ll	ul	p.val	exp.b	ll.exp.b	ul.exp.b	percent	CI
1	religious	-0.034	0.014	-2.380	-0.061	-0.006	0.023	0.967	0.940	0.994	-3.315	95 %
2	minority	-0.221	0.156	-1.418	-0.526	0.084	0.146	0.802	0.591	1.088	-19.807	95 %
3	female	-0.983	0.086	-11.454	-1.151	-0.814	0.000	0.374	0.316	0.443	-62.567	95 %
4	age	-0.005	0.002	-2.335	-0.010	-0.001	0.026	0.995	0.990	0.999	-0.548	95 %
5	emp1	0.486	0.226	2.150	0.043	0.929	0.040	1.626	1.044	2.532	62.576	95 %
6	emp2	0.776	0.245	3.165	0.296	1.257	0.003	2.173	1.344	3.515	117.324	95 %

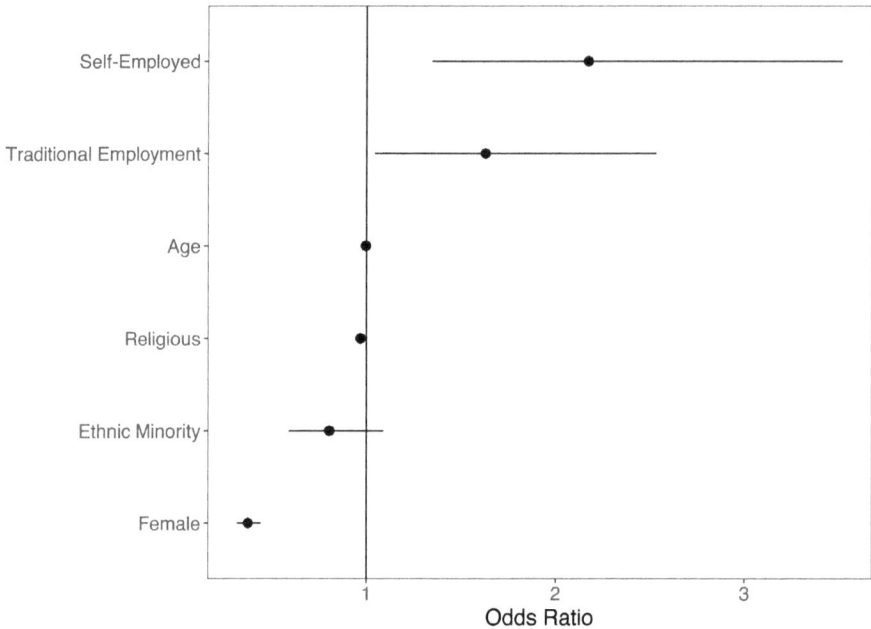

FIGURE 7.4
Plot of odds ratios from ORM.

Like with the log-odds coefficients, we can easily plot these odds ratios using ggplot, resulting in Figure 7.4:

```
ggplot(m1out,aes(y=reorder(variables,exp.b),x=exp.b,xmin=ll.
exp.b,xmax=ul.exp.b)) +
  theme_bw() + geom_pointrange()+geom_vline(xintercept=1)+
  labs(x="Odds Ratio",y="")+scale_y_discrete(labels=c("female"
="Female",
"emp1"="Traditional Employment",
"age"="Age",
"religious"="Religious",
"minority"="Ethnic Minority",
"emp2"="Self-Employed"))
```

As shown in the figure, the same pattern of coefficients emerges as with the log-odds coefficients. This should be unsurprising given that the odds ratio is simply a transformation of the log-odds coefficients. Nevertheless, odds ratios provide more substantively meaningful metrics than a latent, unobserved variable with no natural substantive metric. The quirk with confidence intervals for odds ratios being asymmetrical because exponents are nonlinear transformations is clearly and visually illustrated with the employment variables, where the left side of the confidence interval is shorter in length than the right side. This is one of the reasons why confidence intervals rather than exponentiated standard errors are reported for odds ratios.

In trying to determine the substantive magnitude of the effects of age and religiosity, we may use larger unit changes like decades for age. Therefore, a decade change in age would be interpreted as being associated with a decrease in the odds of feeling safer at night by a factor of .951 ($e^{-0.005 \times 10} = .951$). This is certainly larger than the effect of a single year change in age, but a 5% change in the odds over the course of a decade is a relatively small effect given that being female is associated with a much larger 63% decrease in the odds of feeling safer at night. The same exercise can be done with religiosity, with say a standard deviation (3.06) instead of a unit change.

Predicted Probabilities

Predicted probabilities and changes in them are preferred because they are substantively the most interpretable. The trade-off with the ORM that was not relevant with the BRM is one of dimensionality. Rather than having one set of predicted probabilities for the outcome, our model predicts the probability of each response option. For our four-category ordinal outcome, it means four times as many results to interpret. The suite of `margins` functions from the `catregs` package simplifies generating and testing for differences in predicted probabilities.

Starting with religiosity as an example, we use `margins.des` to create a design matrix for `margins.dat` to calculate the predictions:

```
> design <- margins.des(ml,ivs=expand.grid(religious=0:10))
> pdat <- margins.dat(ml,design)
> pdat
```

	religious	minority	female	age	emp1	emp2	walk.alone.dark	prob	se	ll	ul
1	0	0.077	0.544	53.146	0.799	0.168	Very unsafe	0.055	0.005	0.045	0.065
1.1	0	0.077	0.544	53.146	0.799	0.168	Unsafe	0.155	0.010	0.136	0.174
1.2	0	0.077	0.544	53.146	0.799	0.168	Safe	0.500	0.011	0.477	0.522
1.3	0	0.077	0.544	53.146	0.799	0.168	Very safe	0.290	0.015	0.261	0.318
2	1	0.077	0.544	53.146	0.799	0.168	Very unsafe	0.057	0.005	0.047	0.067
2.1	1	0.077	0.544	53.146	0.799	0.168	Unsafe	0.159	0.009	0.142	0.177
2.2	1	0.077	0.544	53.146	0.799	0.168	Safe	0.501	0.011	0.479	0.523
2.3	1	0.077	0.544	53.146	0.799	0.168	Very safe	0.283	0.012	0.258	0.307
3	2	0.077	0.544	53.146	0.799	0.168	Very unsafe	0.059	0.005	0.049	0.068
3.1	2	0.077	0.544	53.146	0.799	0.168	Unsafe	0.163	0.008	0.147	0.180
3.2	2	0.077	0.544	53.146	0.799	0.168	Safe	0.502	0.011	0.480	0.524
3.3	2	0.077	0.544	53.146	0.799	0.168	Very safe	0.276	0.011	0.255	0.297
4	3	0.077	0.544	53.146	0.799	0.168	Very unsafe	0.060	0.005	0.051	0.070
4.1	3	0.077	0.544	53.146	0.799	0.168	Unsafe	0.167	0.008	0.151	0.183
4.2	3	0.077	0.544	53.146	0.799	0.168	Safe	0.503	0.011	0.481	0.525
4.3	3	0.077	0.544	53.146	0.799	0.168	Very safe	0.269	0.010	0.250	0.289
5	4	0.077	0.544	53.146	0.799	0.168	Very unsafe	0.062	0.005	0.053	0.072
5.1	4	0.077	0.544	53.146	0.799	0.168	Unsafe	0.171	0.008	0.155	0.187
5.2	4	0.077	0.544	53.146	0.799	0.168	Safe	0.504	0.011	0.481	0.526
5.3	4	0.077	0.544	53.146	0.799	0.168	Very safe	0.263	0.010	0.244	0.282
6	5	0.077	0.544	53.146	0.799	0.168	Very unsafe	0.064	0.005	0.054	0.075
6.1	5	0.077	0.544	53.146	0.799	0.168	Unsafe	0.175	0.009	0.158	0.192
6.2	5	0.077	0.544	53.146	0.799	0.168	Safe	0.504	0.011	0.482	0.526
6.3	5	0.077	0.544	53.146	0.799	0.168	Very safe	0.256	0.010	0.236	0.276

7	0.077	0.544	53.146	0.799	0.168	6	Very unsafe	0.066	0.006	0.055	0.078
7.1	0.077	0.544	53.146	0.799	0.168	6	Unsafe	0.179	0.009	0.161	0.198
7.2	0.077	0.544	53.146	0.799	0.168	6	Safe	0.504	0.011	0.482	0.526
7.3	0.077	0.544	53.146	0.799	0.168	6	Very safe	0.250	0.011	0.228	0.272
8	0.077	0.544	53.146	0.799	0.168	7	Very unsafe	0.069	0.006	0.056	0.081
8.1	0.077	0.544	53.146	0.799	0.168	7	Unsafe	0.184	0.010	0.163	0.204
8.2	0.077	0.544	53.146	0.799	0.168	7	Safe	0.504	0.011	0.482	0.526
8.3	0.077	0.544	53.146	0.799	0.168	7	Very safe	0.244	0.013	0.218	0.269
9	0.077	0.544	53.146	0.799	0.168	8	Very unsafe	0.071	0.007	0.057	0.084
9.1	0.077	0.544	53.146	0.799	0.168	8	Unsafe	0.188	0.012	0.165	0.211
9.2	0.077	0.544	53.146	0.799	0.168	8	Safe	0.504	0.011	0.482	0.526
9.3	0.077	0.544	53.146	0.799	0.168	8	Very safe	0.238	0.015	0.209	0.266
10	0.077	0.544	53.146	0.799	0.168	9	Very unsafe	0.073	0.008	0.058	0.088
10.1	0.077	0.544	53.146	0.799	0.168	9	Unsafe	0.192	0.013	0.166	0.218
10.2	0.077	0.544	53.146	0.799	0.168	9	Safe	0.503	0.011	0.481	0.526
10.3	0.077	0.544	53.146	0.799	0.168	9	Very safe	0.231	0.016	0.199	0.263
11	0.077	0.544	53.146	0.799	0.168	10	Very unsafe	0.075	0.009	0.059	0.092
11.1	0.077	0.544	53.146	0.799	0.168	10	Unsafe	0.196	0.015	0.168	0.225
11.2	0.077	0.544	53.146	0.799	0.168	10	Safe	0.503	0.012	0.480	0.525
11.3	0.077	0.544	53.146	0.799	0.168	10	Very safe	0.226	0.018	0.190	0.261

Although the call is similar to our call in the BRM, the resulting output is much more unwieldy. After running `margins.dat`, we can see that there are four times as much output for the 11 possible values of religiosity as a comparable call would produce in the BRM. This is because we now have predicted probabilities for a respondent selecting each of the four possible response options. Although we can interpret the raw predicted probabilities, giving the object returned by `margins.dat` to a plotting function, we can create a predicted probability plot like Figure 7.5:

```
ggplot(pdat,aes(x=religious,y=prob,ymin=ll,ymax=ul,group=dv,li
netype=dv,color=dv)) +
  theme_bw() + geom_line() + geom_point() + geom_
ribbon(alpha=.1) + labs(x="Religious",y="Predicted Probability
",color="",linetype="") +
  scale_color_manual(values=natparks.pals("Glacier")) +
  theme(legend.position="bottom")
```

Looking at the figure, we can clearly see that higher levels of religiosity is associated with a significant decrease in the probability of feeling "very safe"

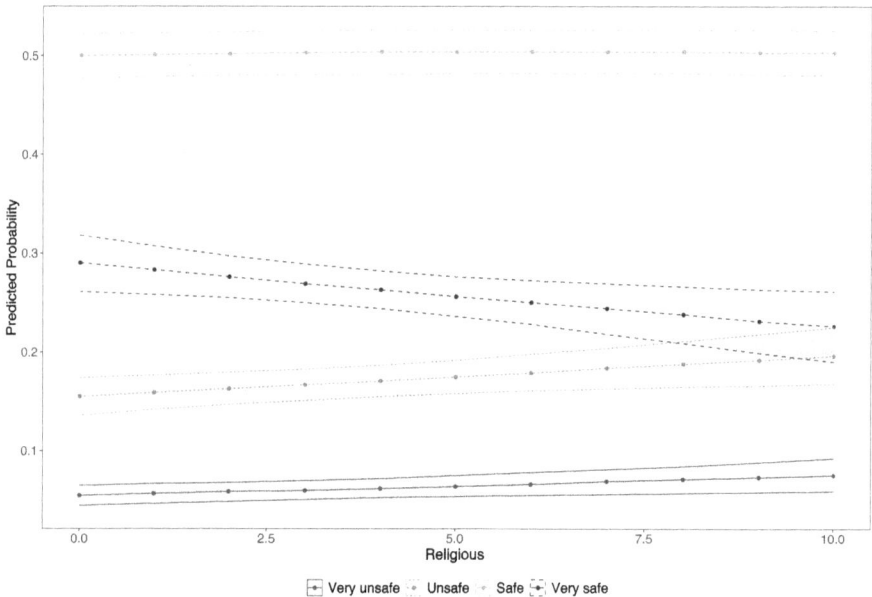

FIGURE 7.5
Predicted probability of feeling safe at night by religiosity.

at night. It has very little effect across the range of the variable on feeling
"safe." Likewise, it is associated with increases in the probability of feeling
"unsafe" and "very unsafe" at night. Rather than break the probabilities up,
another approach to plotting multiple probabilities is a column graph, with
color proportional to the probabilities. Figure 7.6 is such a column graph. This
more clearly illustrates, to us at least, the effect of religiousness on feeling
safe walking alone at night. This graph shows a slight increase in the probability of feeling "very unsafe," another increase in the probability of feeling
"unsafe," little change in the probability of feeling "safe," and a decrease in
the probability of reporting feeling "very safe."

```
ggplot(pdat,aes(x=religious,y=prob,fill=dv)) + theme_bw() +
  geom_col(width=.95) +labs(x="Religion",y="Probability",
fill="")+
  scale_fill_manual(values=natparks.pals("Yellowstone")) +
  theme(legend.position="bottom")
```

The same techniques can be used to interpret dummy or indicator-indepen-
dent variables such as being female:

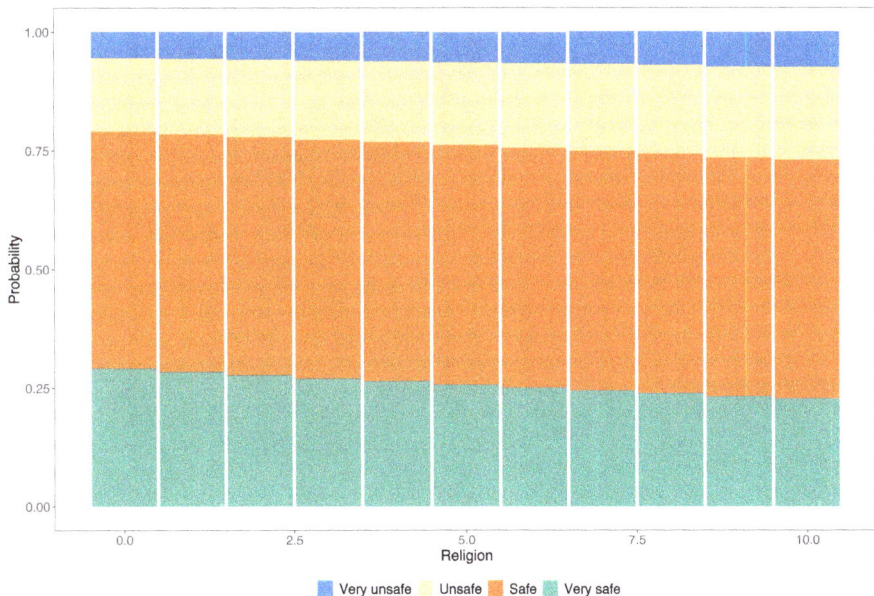

FIGURE 7.6
Predicted probability of each response category of feeling safe at night by religiosity.

```
> design <- margins.des(m1,ivs=expand.grid(female=c(0,1)))
> pdat <- margins.dat(m1,design)
> pdat
```

	female	religious	minority	age	emp1	emp2	walk.alone.dark	prob	se	ll	ul
1	0	3.602	0.077	53.146	0.799	0.168	Very unsafe	0.037	0.004	0.030	0.044
1.1	0	3.602	0.077	53.146	0.799	0.168	Unsafe	0.113	0.008	0.098	0.127
1.2	0	3.602	0.077	53.146	0.799	0.168	Safe	0.469	0.012	0.446	0.492
1.3	0	3.602	0.077	53.146	0.799	0.168	Very safe	0.381	0.015	0.352	0.411
2	1	3.602	0.077	53.146	0.799	0.168	Very unsafe	0.093	0.008	0.078	0.108
2.1	1	3.602	0.077	53.146	0.799	0.168	Unsafe	0.227	0.011	0.205	0.249
2.2	1	3.602	0.077	53.146	0.799	0.168	Safe	0.492	0.011	0.470	0.515
2.3	1	3.602	0.077	53.146	0.799	0.168	Very safe	0.188	0.010	0.167	0.208

Here, the output is more interpretable, but the returned object can also be given to a plotting function to create a predicted probability plot like Figure 7.7. Because we are plotting a categorical independent variable, we use bar graphs instead of lines like we did with the continuous religiosity variable.

As shown in both the figure and raw output, female respondents have a significantly higher predicted probability of feeling "very unsafe" compared to their male counterparts, all else equal (female respondents have a predicted probability of feeling very unsafe of .093 compared to male respondents' .037). We can use the first.diff.fitted function to test if this and other differences are statistically significant:

```
first.diff.fitted(m1,design,compare=c(2,1))
   term     est std.error statistic p.value      ll      ul
1    b1   0.056     0.006     9.238   0.000   0.044   0.068
2    b2   0.114     0.010    10.933   0.000   0.094   0.135
3    b3   0.024     0.008     3.136   0.002   0.009   0.038
4    b4  -0.194     0.017   -11.480   0.000  -0.227  -0.161
```

Although we can clearly gauge statistical significance from the nonoverlapping confidence intervals across most of the values of the dependent

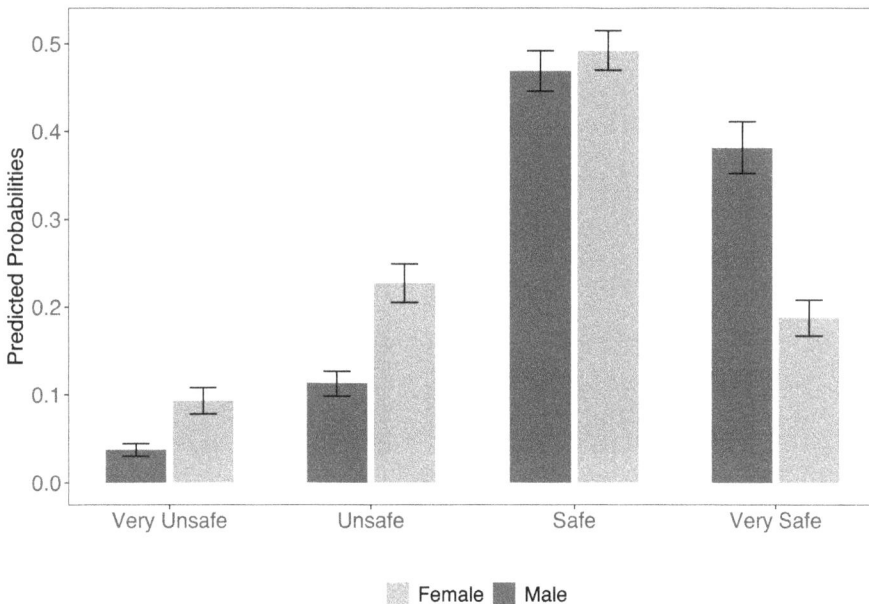

FIGURE 7.7
Predicted probability of feeling safe at night by respondent sex.

variables, we see a nice demonstration of a case of overlapping confidence intervals not necessarily meaning that the difference between two groups is nonsignificant (Schenker and Gentleman 2001). As shown in our test of first differences, although the difference between male and female respondents is small (.024), that difference is nevertheless a statistically significant difference ($p<.01$). This means that across *all* levels of safety, female respondents feel significantly less safe walking at night than her male counterpart. We can also show this gap across the range of the outcome using a column graph or cumulative probability graph like those in Figures 7.8 and 7.9.

Average and Conditional Marginal Effects

Our interpretations up to this point rely on marginal effects at the mean (MEM), but we can do the same using average marginal effects (AME), and perhaps we want to examine the conditional marginal effects. We walk through these alternatives below. Recall from the previous chapters that the `marginaleffects` function can generate the average marginal effect of

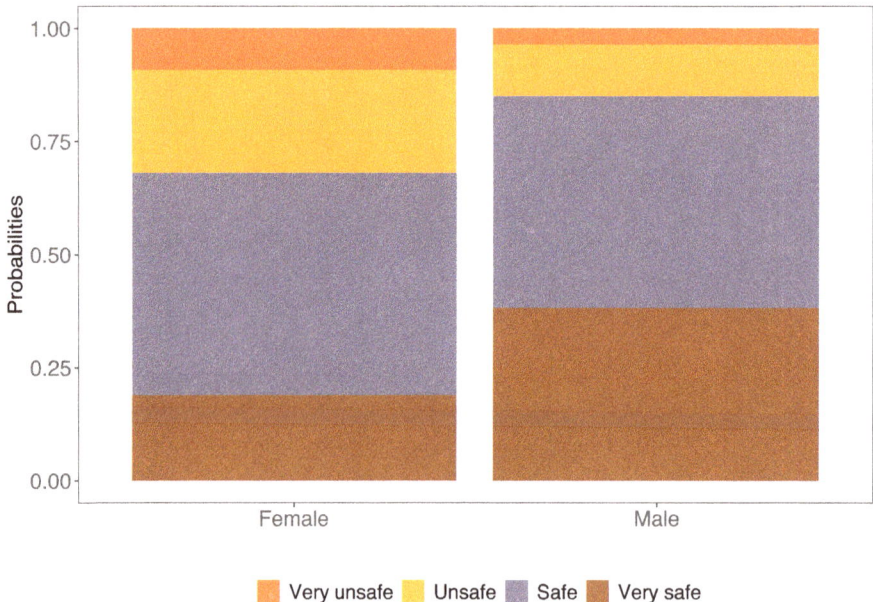

FIGURE 7.8
Predicted probability of feeling safe at night by respondent sex.

some or all variables in a model. Below, we compute the AME of our independent variables and plot them:

```
summary(marginaleffects(m1,variables="female"))
          Group    Term   Effect Std. Error z value  Pr(>|z|)      2.5 %    97.5 %
1 Very unsafe female  0.06246   0.007114   8.780  < 2e-16   0.048519   0.07641
2      Unsafe female  0.11010   0.009729  11.317  < 2e-16   0.091032   0.12917
3        Safe female  0.01332   0.006207   2.146 0.031899   0.001153   0.02549
4   Very safe female -0.18588   0.015343 -12.115  < 2e-16  -0.215956  -0.15581
```

We can see from the AME output and Figure 7.10 that the AME of being female is similar to the MEM calculated above, with female respondents being significantly more likely to report feeling "very unsafe," "unsafe," and "safe" but significantly less likely to report feeling "very safe" relative to their male counterparts. We can also examine the intersection of respondent sex and religiosity using conditional AME. As an example, we calculate the AME of being female at two representative values of religiosity, 1 and 6 on the 10-point scale. To do this, we specify for which variable we want to calculate the AME using the "variables" option and set the value of the religiosity

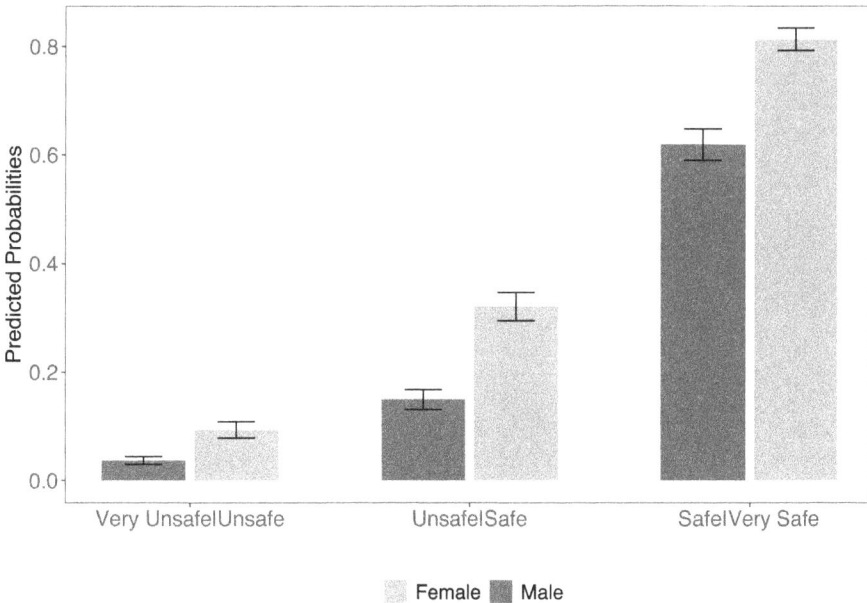

FIGURE 7.9
Cumulative probability of feeling safe at night by respondent sex.

variable using the "newdata=datagrid" option. The full call is below and produces the following output:

```
summary(marginaleffects(m1,variables="female",newdata=datagrid(religious=1)))
          Group    Term    Effect Std. Error z value    Pr(>|z|)     2.5 %    97.5 %
1 Very unsafe female   0.05260   0.006158    8.541 < 2.22e-16   0.04053   0.06467
2       Unsafe female   0.11379   0.011175   10.182 < 2.22e-16   0.09188   0.13569
3         Safe female   0.03293   0.010228    3.220  0.0012835   0.01288   0.05298
4   Very safe female  -0.19931   0.017180  -11.601 < 2.22e-16  -0.23299  -0.16564

summary(marginaleffects(m1,variables="female",newdata=datagrid(religious=6)))
Model type:  polr
Prediction type:  probs
          Group    Term     Effect Std. Error  z value Pr(>|z|)     2.5 %    97.5 %
1 Very unsafe female  0.060979    0.00655    9.3102  < 2e-16   0.04814   0.07382
2       Unsafe female  0.121219    0.01146   10.5744  < 2e-16   0.09875   0.14369
3         Safe female  0.002005    0.00973    0.2061  0.83672  -0.01707   0.02108
4   Very safe female -0.184203    0.01679  -10.9682  < 2e-16  -0.21712  -0.15129
```

Here, we can see that the AME of female is generally larger for lower values of the dependent variable for more religious respondents. In contrast, they are larger for higher values of the dependent variable for less religious respondents. In other words, when it comes to feeling unsafe (very unsafe or unsafe), the difference between female and male respondents are larger among more religious

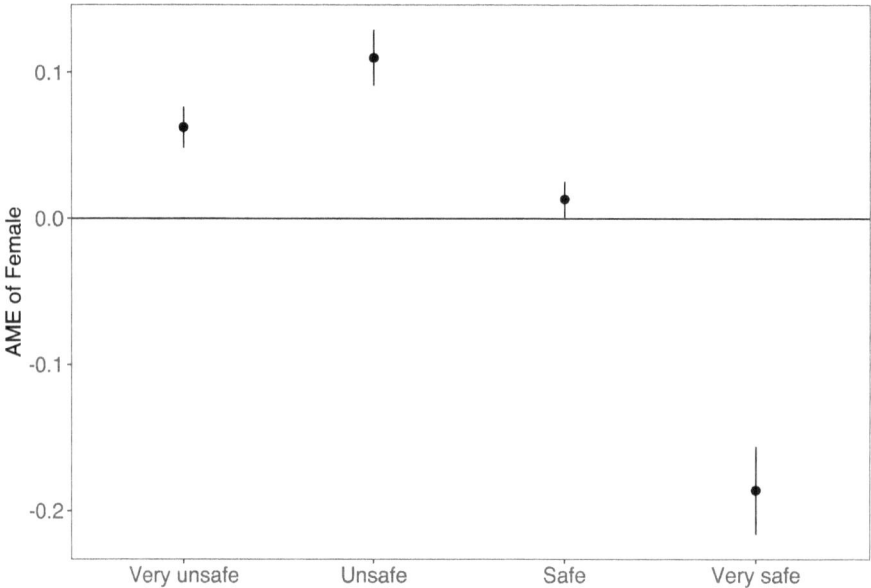

FIGURE 7.10
Average marginal effect of female on feeling safe walking alone at night.

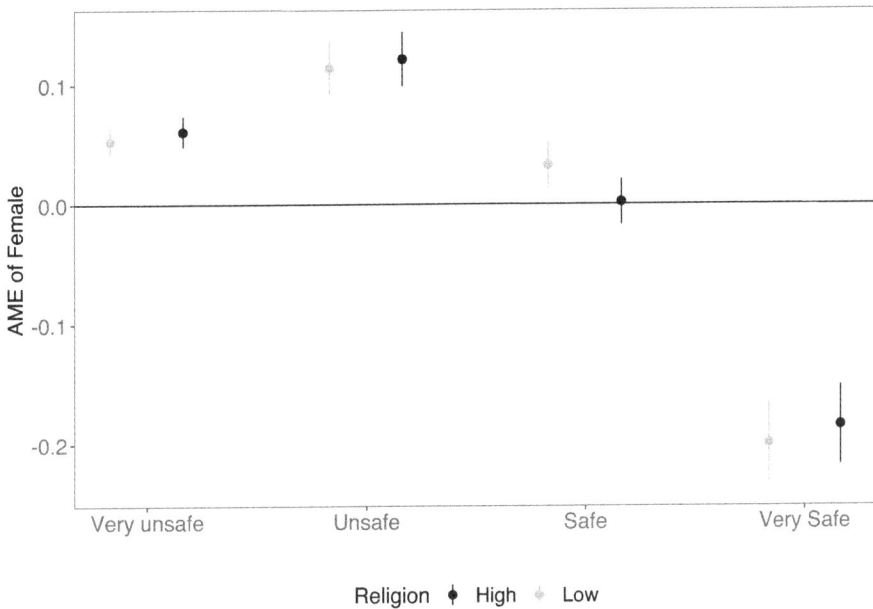

FIGURE 7.11
Conditional average marginal effect of female on feeling safe walking alone at night. AMEs are conditional on religion being low (=1) or high (=6).

respondents than among less religious respondents. In contrast, the differences between female and male respondents are smaller among less religious respondents. This conditional effect can be visually demonstrated in Figure 7.11.

The Parallel Regression Assumption

As noted above, an important assumption of the ORM is that the coefficients for all categories of m are the same. This is called the parallel regression assumption, or for the ordered logit the proportional odds assumption, because we only estimated one structural model for all values of the outcome. Indeed, the only thing that distinguishes what predicts reporting feeling "Very unsafe" versus "Very safe" on either extreme of our example is the cut point between these categories. If we were to plot the cumulative probability of our Eq. 7.4 using some arbitrary values for x, we would get Figure 10. This figure illustrates that each of these curves are the same, just shifted based on the values of our estimated cut points. These regressions, in other words, are parallel to one another.

Is this a reasonable assumption? Maybe. Theoretically, it is a question of whether a researcher believes that a single process determines different levels of the ordinal outcome in the latent-variable formulation of the model. Do we have good theoretical reason to expect that our independent variables predict some underlying feeling of safety at night and these ordinal manifestations of this underlying feeling are just where people fall along this latent variable? Or, do we think that the effect of say, being female, depends on whether a respondent is feeling "Safe" versus "Very safe" or "Unsafe" versus "Very unsafe?" Empirically, this is an assumption that is testable. Practically, it is unclear what to do when the assumption fails (Long and Freese 2006). We walk through the Brant (1990) test for examining this assumption and potential options when the assumption fails.

The Brant (1990) test is a Wald test of individual BRMs at various cut points. In essence, the test estimates $J - 1$ BRMs for each of the cut points and then tests whether the coefficients corresponding to each predictor are significantly different from one another with the null hypothesis being no difference. In other words, if there is one underlying process determining the values of our latent variable, it should not matter if we are binarizing the variable at "Very unsafe" versus everything else or "Safe or above" versus "Unsafe or below." The values of the coefficients should be similar. The brant function in the `brant` package (Schlegel and Steenbergen 2020) estimates and reports results from this test for our example model:

```
brant
------------------------------------------------
Test for      X2      df      probability
------------------------------------------------
Omnibus       23.97   12      0.02
religious     1.89    2       0.39
minority      1.72    2       0.42
female        4.48    2       0.11
age           4.64    2       0.1
emp1          0.49    2       0.78
emp2          4.2     2       0.12
------------------------------------------------

H0: Parallel Regression Assumption holds
```

The first line is an "omnibus" test, referring to the joint test of significance for all coefficients in the model. Based on this result, our model significantly violates the parallel regression assumption. However, the Brant test also tests each individual coefficient across cut points. Here, we do not find any main culprit per se. One way to interpret this discrepancy is to say that none of the individual coefficients are significantly different across cut points but collectively they are different enough that we are confident that the overall model violates the assumption.

What can we do in this case? There are three main options: (1) relaxing the parallel regression assumption, (2) abandoning the assumption altogether, or (3) binarizing the outcome. It is unclear which option produces the best model (Long and Freese 2006), but theory and disciplinary norms should

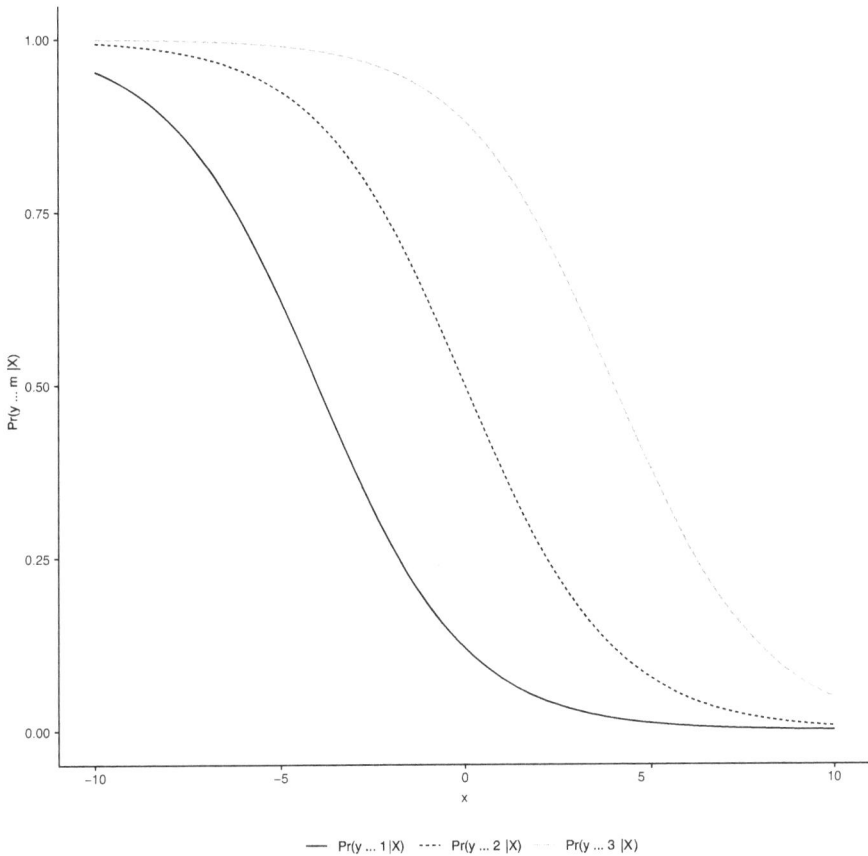

FIGURE 7.12
Illustration of the parallel regression assumption.

guide the decision. We walk through the first option in this chapter. The second and third options rely on models covered in Chapters 8 and 5, respectively, so we do not cover them here.

Partial Proportional Odds Models

If one had theoretical reasons to prefer an ordinal model, but the parallel regression assumption is violated, the most straightforward solution is to relax that assumption. This results in a "partial proportional odds" model (Peterson and Harrell 1990) for an ordered logit. The logic can be applied for ordered probit as well, but software to implement a "somewhat parallel regression" model is lacking. A partial proportional odds model fits a multinomial logit model (covered in Chapter 8) for variables that violate the parallel regression assumption but keeps the ordered logit model for variables that do not violate this assumption (Williams 2005). This relaxes the assumption without abandoning it altogether. In essence, it says there is one underlying process that determines values on the latent outcome, but some variables have different processes that are more complicated.

Despite its intuitive appeal, it has two noteworthy problems that need to be checked. First, if every variable violates the assumption, the partial proportional odds model is just a multinomial logit and is not even an ordinal model any longer (Clogg and Shihadeh 1994). Second, and perhaps more consequential, because the model is concatenating two separate types of models together, predicted probabilities can sometimes go negative to make the models jointly converge (Williams 2005). Given neither of these problems are present, the partial proportional odds model can be estimated using the vglm function from the VGAM package (Yee, Stoklosa, and Huggins 2015). Using the following call, we get results from this model:

```
m2 <- vglm(dv ~ religious + minority  + female + age + emp1 + emp2,
           cumulative(parallel = FALSE ~ age + female, reverse =
FALSE),
           data = dat)
summary(m2)

Call:
vglm(formula = dv ~ religious + minority + female + age + emp1 +
    emp2, family = cumulative(parallel = FALSE ~ age + female,
    reverse = FALSE), data = dat)

Coefficients:
                Estimate Std. Error z value Pr(>|z|)
(Intercept):1 -3.962146    0.390133 -10.156  < 2e-16 ***
(Intercept):2 -1.729878    0.271020  -6.383 1.74e-10 ***
(Intercept):3  0.722311    0.263105   2.745  0.00604 **
religious      0.032919    0.014053   2.343  0.01915 *
minority       0.228745    0.156254   1.464  0.14321
female:1       1.363921    0.216448   6.301 2.95e-10 ***
female:2       1.035135    0.111646   9.272  < 2e-16 ***
female:3       0.921326    0.100440   9.173  < 2e-16 ***
age:1          0.014590    0.004747   3.074  0.00212 **
age:2          0.006434    0.002918   2.205  0.02745 *
age:3          0.003295    0.002795   1.179  0.23851
emp1          -0.498082    0.228123  -2.183  0.02901 *
emp2          -0.782326    0.246725  -3.171  0.00152 **
---
Signif. codes:  0 '***' 0.001 '**' 0.01 '*' 0.05 '.' 0.1 ' ' 1

Names of linear predictors: logitlink(P[Y<=1]),
logitlink(P[Y<=2]), logitlink(P[Y<=3])

Residual deviance: 4997.049 on 6476 degrees of freedom
```

```
Log-likelihood: -2498.524 on 6476 degrees of freedom

Number of Fisher scoring iterations: 5
```

Although none of our individual variables violate the assumption, we estimated the partial proportional odds model for illustration purposes with `female` and `age` relaxed because they have the lowest *p*-values. The `parallel = FALSE ~` option lets us specify for which variables to relax the assumption. Having fit the model, we now see that the output reports cut point–specific coefficients for these two variables. This naturally raises the question (especially because neither of these variables violated the assumption) of whether the extra degrees of freedom spent on estimating these additional coefficients improves the fit of the models. We can compare the AIC and BIC for this model compared to our original ordinal model:

```
> AIC(m1);AIC(m2)
[1] 5024.31
[1] 5023.049
> BIC(m1);BIC(m2)
[1] 5075.423
[1] 5096.879
```

Here, we get mixed results with the AIC preferring our original model (AIC=5024.31 vs. 5023.05) while the BIC provides "very strong" evidence that the partial proportional odds model better fits the data (BIC=5096.88 vs. 5075.42; Raftery 1995).

A more substantive comparison is to examine whether the relaxation of this assumption substantively impacts our conclusions. Figures 7.13 and 7.14 show a side-by-side comparison of the effects of age for our new model (shown on the left) with our original model (shown on the right). As shown in these figures, the effects of both variables are substantively similar. The effect of age on the top category of "Very safe" is noticeably weaker (flatter slope) than when we relax the parallel regression assumption, while the effect of age on "Safe" is stronger. This suggests that the assumption may unnecessarily overstate the effects of age toward the higher extreme of the ordinal scale. Nevertheless, the conclusions are similar between the two models. The effects of respondent sex are substantively similar between the two models.

Nominal and Binary Models

Given that the partial proportional odds model produces substantively similar results to our original model, we can try to address the failed assumption test in two other ways. We could completely abandon the parallel regression assumption and fit a multinomial logit, which treats that problem as a discrete choice problem. In other words, there is no underlying process

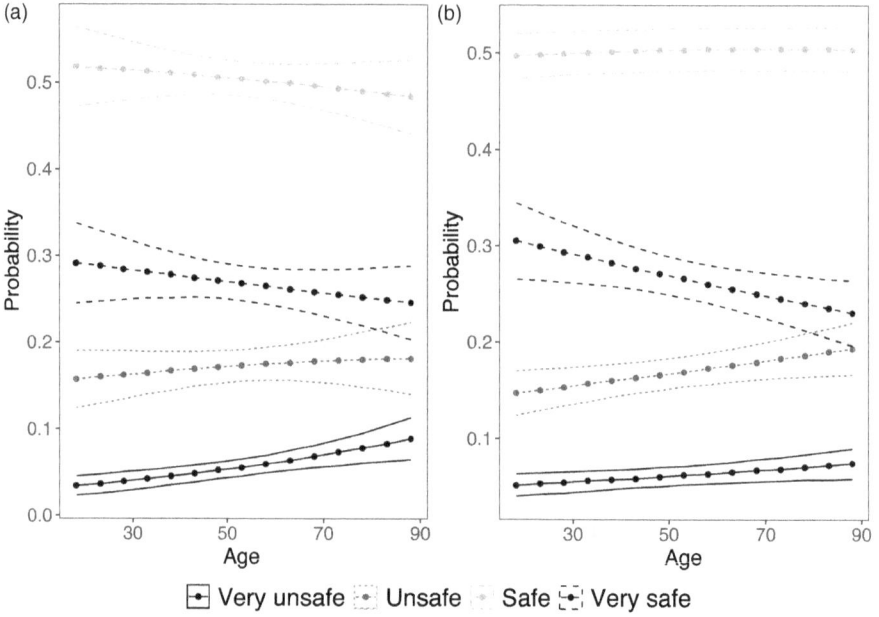

FIGURE 7.13
Comparison of predicted probability by age.

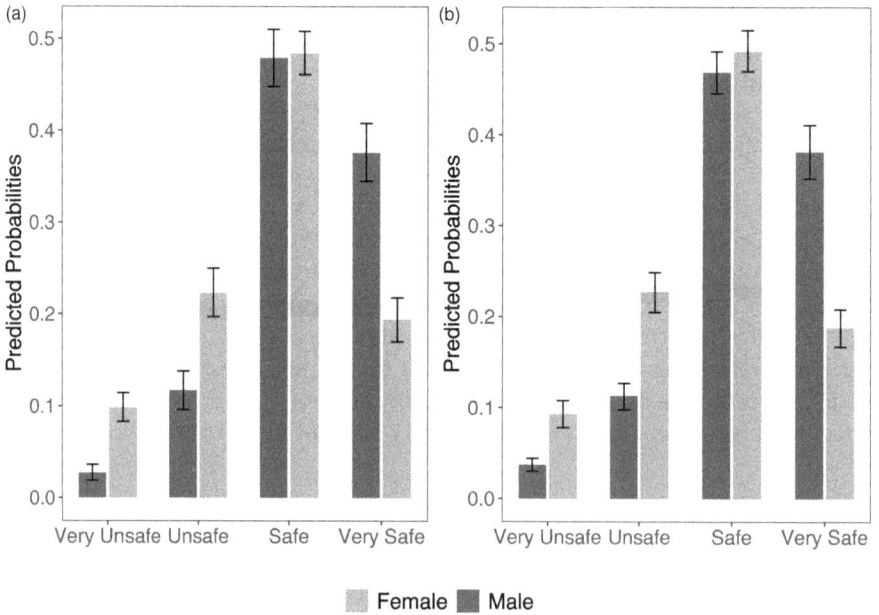

FIGURE 7.14
Comparison of predicted probability by respondent sex.

that determines whether someone reports "Very unsafe," "Unsafe," "Safe," or "Very unsafe." Instead, when responding to the survey question asking how safe they feel walking alone at night, they are making pairwise choices between each of these options and selecting the best one. This model throws out the ordinality of the outcome variable – and thus the parallel regression assumption – and is the topic of the next chapter. Finally, we could instead simplify the problem and binarize the outcome as we did in Chapter 5 (Snijders and Bosker 2011). There is no single agreed upon solution and, all else equal, we suggest presenting the most theoretically defensible model and/or the best fitting model with the alternatives as sensitivity checks.

Note

1. There certainly are examples where specific manifestations of an ordinal outcome are important to examine. Anyone who has worked in customer service can probably relate to a customer satisfaction survey where "Super extremely satisfied" is the only acceptable outcome. Thus, the outcome itself is important not just an underlying propensity for satisfaction. However, these are relatively unique circumstances and reflect deeper survey design issues.

8

Regression for Nominal Outcomes

Nominal outcomes are categorical outcomes that are unordered. Categories like marital status, occupation, and political affiliation, among others, have no inherent ordering to them and are thus nominal outcomes. However, ordinal outcomes can also be treated as nominal for analytical purposes. The key trade-off is one of efficiency. If the parallel regression assumption holds, an ordinal outcome can be modeled using one set of coefficients. However, if the assumption does not hold or if the researcher, for theoretical or practical reasons, chooses to ignore the ordinality of the outcome, models designed for nominal outcomes are good alternatives. We cover these models in this chapter, focusing on the multinomial logit model.

The Multinomial Regression Model and Its Assumptions

A multinomial regression model (MRM) can be thought of as a series of pairwise BRMs. Because we are not assuming any ordering to the categories, it follows that nothing ties the categories together. Thus, we simultaneously fit pairwise comparisons for each category of the outcome. To demonstrate the relationship between the MRM and the ORM, we will continue to use the feeling safe walking at night variable used in previous chapters. Suppose that for the four categories of this outcome, we wanted to know the probability that a respondent selects each of the response categories. As with our previous derivations, there are two common ways of arriving at the same model: the statistical approach and the latent variable approach. We begin with the latent variable approach because, unlike the ORM, the MRM is less often discussed as a latent variable model.

In the latent variable approach to the MRM, there is an underlying propensity for each response category:

$$y_i^{1*} = \mathbf{Xb}_1 + \varepsilon_1 \tag{8.1}$$

$$y_i^{2*} = \mathbf{Xb}_2 + \varepsilon_2$$

DOI: 10.1201/9781003029847-8

$$y_i^{3*} = \mathbf{Xb}_3 + \varepsilon_3$$

$$y_i^{4*} = \mathbf{Xb}_4 + \varepsilon_4$$

where the distribution of ε is assumed to be normal for the multinomial probit and an extreme value distribution for the multinomial logit. Note that each category has its own coefficient and error term, unlike the ORM, which assumes the same underlying propensity. Given this, the value of y_i is the category with the highest latent propensity:

$$y_i = \begin{cases} 1 & \text{if} & y_i^{1*} > y_i^{2*} > \ldots > y_i^{J*} \\ 2 & \text{if} & y_i^{2*} > y_i^{1*} > \ldots > y_i^{J*} \\ 3 & \text{if} & y_i^{3*} > y_i^{1*} > \ldots > y_i^{J*} \\ \vdots & & \vdots \\ J & \text{if} & y_i^{J*} > y_i^{1*} > \ldots > y_i^{J-1*} \end{cases} \tag{8.2}$$

As is the case with the ORM, the BRM is also a special case of the MRM where there are only two categories of the outcome. Using the statistical approach, the MRM is a simultaneously fit series of BRMs for each pairwise comparison of the categories. This results in the following multinomial formulation:

$$\ln \frac{\Pr(y = m \mid x)}{\Pr(y = b \mid x)} = \mathbf{Xb}_{m|b} + \varepsilon_{m|b} \text{ for } m = 1 \text{ to } J \tag{8.3}$$

where b is a base outcome or reference category. To get the predicted probabilities of any given category, we exponentiate each side and sum across the pairwise comparisons:

$$\Pr(y = m \mid x) = \exp\left(\mathbf{Xb}_{m|b} + \varepsilon_{m|b}\right) \times \Pr(y = b \mid x) = \frac{\exp\left(\mathbf{Xb}_{m|b} + \varepsilon_{m|b}\right)}{\sum_{j=1}^{J} \exp\left(\mathbf{Xb}_{j|b} + \varepsilon_{j|b}\right)}$$

$$\tag{8.4}$$

Due to the nature of divisions of logged values, a few things make the estimation of the multinomial logit more efficient. First, because $\ln\dfrac{\Pr(y = m \mid x)}{\Pr(y = b \mid x)} = \ln\Pr(y = m \mid x) - \ln\Pr(y = b \mid x)$, we only need to estimate $J - 1$ pairwise comparisons. That is, the final comparison does not need to be estimated; instead it is inferred due to this constraint. Second, because $\ln\dfrac{\Pr(y = b \mid x)}{\Pr(y = b \mid x)} = \ln 1 = 0$, the predicted probability of each category of the outcome is the same regardless of which base category is used to estimate the equations. Because of this, we can easily switch base categories for reporting purposes. Third, because the MRM is a pairwise comparison of all categories, mathematically, it ignores the other categories for each pairwise comparison. This is known as the independence of irrelevant alternatives assumption. We end this chapter discussing this assumption in more detail, including tests for violations of this assumption.

Estimating the MRM

Estimating the MRM in R requires either the nnet (Ripley, Venables, and Ripley 2016) or mlogit package (Croissant 2020). Because the mlogit package is designed for a broader class of choice models than the MRM, it requires that the data be reshaped whereas the nnet package does not. Therefore, we use the multinom function from the nnet package instead of the mlogit function. To estimate our model, we use the following call:

```
m1<-multinom(walk.alone.dark ~ education + religious +
minority  + female + age + selfemp + unemp,data=X)

summary(m1)

Call:
multinom(formula = walk.alone.dark ~ education + religious +
    minority + female + age + selfemp + unemp, data = X)
```

```
Coefficients:
          (Intercept) education   religious    minority     female          age     selfemp     unemp
Unsafe     0.07938611  0.1432192 -0.02355649  0.19616199 -0.8369336 -0.003171997 -0.78919314 0.5791115
Safe       0.25252911  0.2272629 -0.02782981 -0.05162756 -1.4802846 -0.002430671 -0.20026941 0.3617297
Very safe -0.68749840  0.2856166 -0.07781572 -0.52431539 -2.1780517 -0.001448733 -0.06556712 0.2822127

Std. Errors:
          (Intercept) education  religious   minority    female         age   selfemp     unemp
Unsafe      0.6494315 0.03534991 0.03431487 0.3946872 0.2500798 0.005882475 0.2931992 0.5314833
Safe        0.6054639 0.03320408 0.03171533 0.3757739 0.2326298 0.005449356 0.2499562 0.5165850
Very safe   0.6321504 0.03451382 0.03389427 0.4102661 0.2413386 0.005799877 0.2617794 0.5731341

Residual Deviance: 4834.852
AIC: 4882.852
```

The output from `multinom` is different from the output from our BRM and ORM models in two major ways. First, highlighting the multiple equations being estimated, coefficients and standard errors are listed across rows rather than the more typical presentation down a column, with each row representing a pairwise comparison. The intercept and coefficients for the "Unsafe" row, for example, presents the estimates for the comparison of "Unsafe" versus the base category. What the base category is depends on how the factor variable is set up during data cleaning. By default, the base category is the first category for the variable. In this case, all rows are pairwise comparisons between the category named and "Very unsafe," the first category for the outcome variable. Second, there are no *p*-values or test statistics reported. Both differences (display as well as missing statistics) can be addressed by calling `list.coef()` after model:

```
> list.coef(ml)
$out
```

	variables	b	SE	z	ll	ul	p.val	exp.b	ll.exp.b	ul.exp.b	percent	CI
1	Unsafe (Intercept)	0.079	0.649	0.122	-1.193	1.352	0.396	1.083	0.303	3.866	8.262	95 %
2	Unsafe education	0.143	0.035	4.051	0.074	0.213	0.000	1.154	1.077	1.237	28.728	95 %
3	Unsafe religious	-0.024	0.034	-0.686	-0.091	0.044	0.315	0.977	0.913	1.045	-49.717	95 %
4	Unsafe minority	0.196	0.395	0.497	-0.577	0.970	0.353	1.217	0.561	2.637	15.398	95 %
5	Unsafe female	-0.837	0.250	-3.347	-1.327	-0.347	0.001	0.433	0.265	0.707	25.516	95 %
6	Unsafe age	-0.003	0.006	-0.539	-0.015	0.008	0.345	0.997	0.985	1.008	33.058	95 %
7	Unsafe selfemp	-0.789	0.293	-2.692	-1.364	-0.215	0.011	0.454	0.256	0.807	-2.328	95 %
8	Unsafe unemp	0.579	0.531	1.090	-0.463	1.621	0.220	1.784	0.630	5.057	-2.745	95 %
9	Safe (Intercept)	0.253	0.605	0.417	-0.934	1.439	0.366	1.287	0.393	4.217	-7.487	95 %
10	Safe education	0.227	0.033	6.844	0.162	0.292	0.000	1.255	1.176	1.340	21.672	95 %
11	Safe religious	-0.028	0.032	-0.877	-0.090	0.034	0.271	0.973	0.914	1.035	-5.032	95 %
12	Safe minority	-0.052	0.376	-0.137	-0.788	0.685	0.395	0.950	0.455	1.984	-40.804	95 %
13	Safe female	-1.480	0.233	-6.363	-1.936	-1.024	0.000	0.228	0.144	0.359	-56.696	95 %
14	Safe age	-0.002	0.005	-0.446	-0.013	0.008	0.361	0.998	0.987	1.008	-77.243	95 %
15	Safe selfemp	-0.200	0.250	-0.801	-0.690	0.290	0.289	0.819	0.501	1.336	-88.674	95 %
16	Safe unemp	0.362	0.517	0.700	-0.651	1.374	0.312	1.436	0.522	3.952	-0.317	95 %
17	Very safe (Intercept)	-0.687	0.632	-1.088	-1.926	0.551	0.221	0.503	0.146	1.736	-0.243	95 %
18	Very safe education	0.286	0.035	8.275	0.218	0.353	0.000	1.331	1.244	1.424	-0.145	95 %
19	Very safe religious	-0.078	0.034	-2.296	-0.144	-0.011	0.029	0.925	0.866	0.989	-54.579	95 %
20	Very safe minority	-0.524	0.410	-1.278	-1.328	0.280	0.176	0.592	0.265	1.323	-18.149	95 %
21	Very safe female	-2.178	0.241	-9.025	-2.651	-1.705	0.000	0.113	0.071	0.182	-6.346	95 %
22	Very safe age	-0.001	0.006	-0.250	-0.013	0.010	0.387	0.999	0.987	1.010	78.445	95 %
23	Very safe selfemp	-0.066	0.262	-0.250	-0.579	0.448	0.387	0.937	0.561	1.564	43.581	95 %
24	Very safe unemp	0.282	0.573	0.492	-0.841	1.406	0.353	1.326	0.431	4.078	32.606	95 %

By calling `list.coef()`, we reshape the output to be displayed more traditionally, but more importantly, we now have z-statistics, confidence intervals, p-values, as well as exponentiated coefficients (odds ratios).

To change the reference category, we can recode or create a new outcome variable that re-levels the factor variable:

```
> X$walk.alone.dark2 <- relevel(as.factor(X$walk.alone.dark),
ref = "Very safe")

> m1<-multinom(walk.alone.dark2 ~ education + religious +
minority  + female + age + selfemp + unemp,data=X)

> summary(m1)
Call:
multinom(formula = walk.alone.dark2 ~ education + religious +
    minority + female + age + selfemp + unemp, data = X)

Coefficients:
(Intercept) education religious minority female age selfemp unemp
```

	(Intercept)	education	religious	minority	female	age	selfemp	unemp
Very unsafe	0.6872838	-0.28560437	0.07783796	0.5243836	2.178056	0.0014473749	0.06567437	-0.28256885
Unsafe	0.7669118	-0.14239995	0.05426293	0.7205211	1.341167	-0.0017235443	-0.72359723	0.29665532
Safe	0.940239	-0.05835194	0.04998635	0.4727022	0.697793	-0.0009826196	-0.13465654	0.07946517

Std. Errors:

	(Intercept)	education	religious	minority	female	age	selfemp	unemp
Very unsafe	0.6321563	0.03451376	0.03389469	0.4102645	0.2413424	0.005799974	0.2617786	0.5731732
Unsafe	0.4058026	0.02085165	0.02405185	0.2722929	0.1458363	0.004092805	0.2131423	0.3894901
Safe	0.2976612	0.01476473	0.01848972	0.2221339	0.1080930	0.003134465	0.1341375	0.3473129

Residual Deviance: 4834.852
AIC: 4882.852

```
> list.coef(m1)
```

	variables	b	SE	z	ll	ul	p.val	exp.b	ll.exp.b	ul.exp.b	percent	CI
1	Very unsafe (Intercept)	0.687	0.632	1.087	-0.552	1.926	0.221	1.988	0.576	6.864	98.831	95%
2	Very unsafe education	-0.286	0.035	-8.275	-0.353	-0.218	0.000	0.752	0.702	0.804	115.311	95%
3	Very unsafe religious	0.078	0.034	2.296	0.011	0.144	0.029	1.081	1.011	1.155	156.004	95%
4	Very unsafe minority	0.524	0.410	1.278	-0.280	1.328	0.176	1.689	0.756	3.775	-24.844	95%
5	Very unsafe female	2.178	0.241	9.025	1.705	2.651	0.000	8.829	5.502	14.169	-13.273	95%
6	Very unsafe age	0.001	0.006	0.250	-0.010	0.013	0.387	1.001	0.990	1.013	-5.668	95%
7	Very unsafe selfemp	0.066	0.262	0.251	-0.447	0.579	0.387	1.068	0.639	1.784	8.095	95%
8	Very unsafe unemp	-0.283	0.573	-0.493	-1.406	0.841	0.353	0.754	0.245	2.318	5.576	95%
9	Unsafe (Intercept)	0.767	0.406	1.890	-0.028	1.562	0.067	2.153	0.972	4.770	5.126	95%
10	Unsafe education	-0.142	0.021	-6.829	-0.183	-0.102	0.000	0.867	0.833	0.903	68.942	95%

11	Unsafe religious	0.054	0.024	2.256	0.007	0.101	0.031	1.056	1.007	1.107	105.550	95 %
12	Unsafe minority	0.721	0.272	2.646	0.187	1.254	0.012	2.056	1.205	3.505	60.432	95 %
13	Unsafe female	1.341	0.146	9.196	1.055	1.627	0.000	3.824	2.873	5.089	782.913	95 %
14	Unsafe age	-0.002	0.004	-0.421	-0.010	0.006	0.365	0.998	0.990	1.006	282.350	95 %
15	Unsafe selfemp	-0.724	0.213	-3.395	-1.141	-0.306	0.001	0.485	0.319	0.737	100.931	95 %
16	Unsafe unemp	0.297	0.389	0.762	-0.467	1.060	0.298	1.345	0.627	2.886	0.145	95 %
17	Safe (Intercept)	0.940	0.298	3.158	0.357	1.523	0.003	2.560	1.428	4.588	-0.172	95 %
18	Safe education	-0.058	0.015	-3.952	-0.087	-0.029	0.000	0.943	0.916	0.971	-0.098	95 %
19	Safe religious	0.050	0.018	2.703	0.014	0.086	0.010	1.051	1.014	1.090	6.788	95 %
20	Safe minority	0.473	0.222	2.128	0.037	0.908	0.041	1.604	1.038	2.480	-51.500	95 %
21	Safe female	0.698	0.108	6.455	0.486	0.910	0.000	2.009	1.626	2.483	-12.598	95 %
22	Safe age	-0.001	0.003	-0.313	-0.007	0.005	0.380	0.999	0.993	1.005	-24.616	95 %
23	Safe selfemp	-0.135	0.134	-1.004	-0.398	0.128	0.241	0.874	0.672	1.137	34.535	95 %
24	Safe unemp	0.079	0.347	0.229	-0.601	0.760	0.389	1.083	0.548	2.139	8.271	95 %

Calling `relevel` using the `ref` option changes the base category for the MRM. As shown in the output, this changes the coefficients reported, and we can call `list.coef()` to report statistics not reported by default. Comparing the "Very unsafe" coefficients reported here to the "Very safe" coefficients reported in the previous example, it should be clear that these coefficients are direct inverses of each other. This is because these comparisons are the same, just in the opposite direction. The other coefficients are different, because the comparison is not inherently the same. Therefore, it is important to clearly specify what the base category is when reporting results from the MRM, even if this decision is of no mathematical importance. It may be important for theoretical or substantive reasons to prefer a given base category over another.

Interpreting the MRM

Interpreting the MRM is complicated because of the sheer number of coefficients being estimated. Perhaps, more cumbersome is the need to specify what the base category is in interpretations involving the coefficients or transformed versions of the coefficients like the odds ratio. Although our preference for all models presented so far is to use predicted probabilities, this suggestion is stronger in the case of the MRM because it removes the clunkiness of base categories inherent in the other interpretation methods. This is because the base category chosen makes no difference to the predicted probability, as it is calculated for each outcome category. Nevertheless, we walk through interpretations for the regression coefficients as well as the odds ratio in case there are compelling reasons to prefer those for specific cases.

Regression Coefficients

Continuing with our example, Figure 8.1 presents coefficients and corresponding confidence intervals. Given the number of coefficients, presenting in graphical form makes it easier to interpret the patterns. Unlike our earlier examples, it does not make as much sense to sort the coefficients by magnitude in this context. That could put coefficients making drastically different pairwise comparisons next to each other. For example, the effect of religion and education comparing "Very unsafe" to "Very safe" are similar to the effects of self-employment comparing "Unsafe" to "Very safe." This was not the case when there was only one (explicit or implicit) base category ("Not safe" in the BRM and less safe in the ORM). One potential solution is to sort coefficients by magnitude within pairwise comparisons, but that makes comparing the effects of the same coefficients across pairwise comparisons

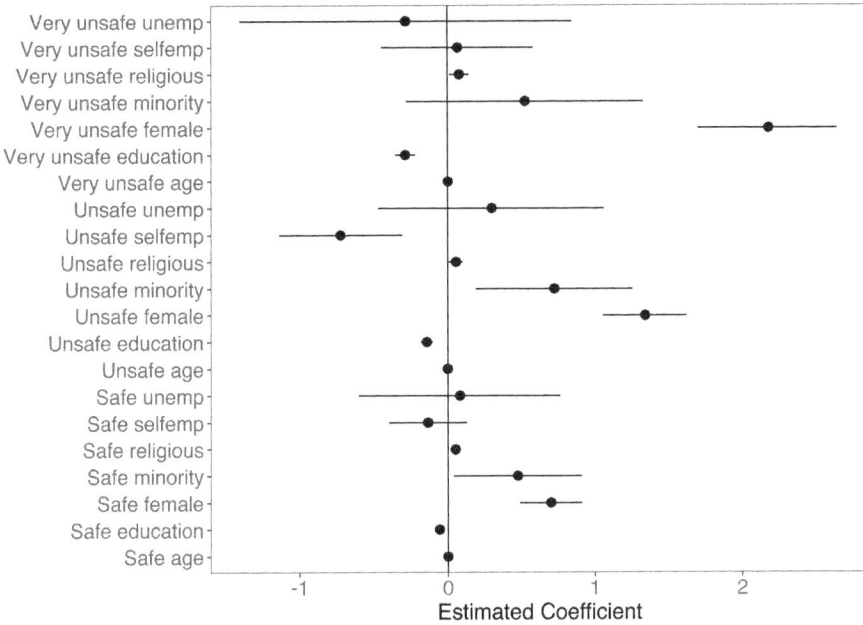

FIGURE 8.1
Coefficient plot and confidence intervals compared to "Very Safe."

more difficult. Ultimately, it is up to the researcher to determine how to balance these concerns. The problems are just more evident in the MRM.

As shown in the figure, a few variables are significantly related to how safe respondents report feeling walking at night. All else equal, self-employment, religiosity, being a racial or ethnic minority, being female, and education each have some significant effects. Being self-employed compared to traditional employment is significantly associated with lower log-odds of reporting feeling "Unsafe" compared to "Very safe." Self-employment is not significantly associated with other response categories compared to "Very safe." Each additional unit of religiosity is associated with a 0.078, 0.054, 0.050 (all $p < .05$) increase in the log-odds for reporting "Very unsafe," "Unsafe," and "Safe," compared to reporting "Very safe," respectively. Religiosity seems to have a monotonic effect on feeling less safe at night. In contrast, being a racial or ethnic minority only has a significant impact on the comparison between the middle categories of "Unsafe" and "Safe" versus "Very safe," with racial or ethnic minorities being less likely to report "Very safe" compared to "Safe" or "Unsafe." Being female significantly associates with a 2.178, 1.341, and 0.698 increase ($p < .001$) in the log-odds of reporting "Very unsafe," "Unsafe," and "Safe" compared to "Very safe," respectively. Like religion, being female has a uniformly negative effect on choosing "Very safe." In contrast, education has a consistent and significant impact on being less likely to report feeling "Very unsafe," "Unsafe," and "Safe" compared to "Very safe" (all $p < .001$).

Odds Ratios

The same comparisons can be made using odds ratios. As shown in Figure 8.2, the variables that are significantly associated with feeling safe at night are the same as those discussed above: self-employment, religiosity, being a racial or ethnic minority, being female, and education. Also shown in the figure is the asymmetrical nature of the confidence intervals for odds ratios compared to the confidence intervals for coefficients.

We could interpret the odds ratios using the factor change (exp.b column of the list.coef output) or percent change (using the percent column). Because we plotted the factor change, we will use the factor change for our interpretation, but the plot can just as easily be based on the percent change. Self-employment compared to traditional employment is associated with a 0.485 and 0.874 times decrease in the odds of reporting feeling "Unsafe" and "Safe" compared to "Very safe," respectively ($p < .05$ across both odds ratios). Each additional unit of reported religiosity is associated with a 1.081, 1.056, 1.051 times increase in the odds of reporting feeling "Safe," "Unsafe," and "Very unsafe," compared to "Very safe," respectively ($p < .05$ across these

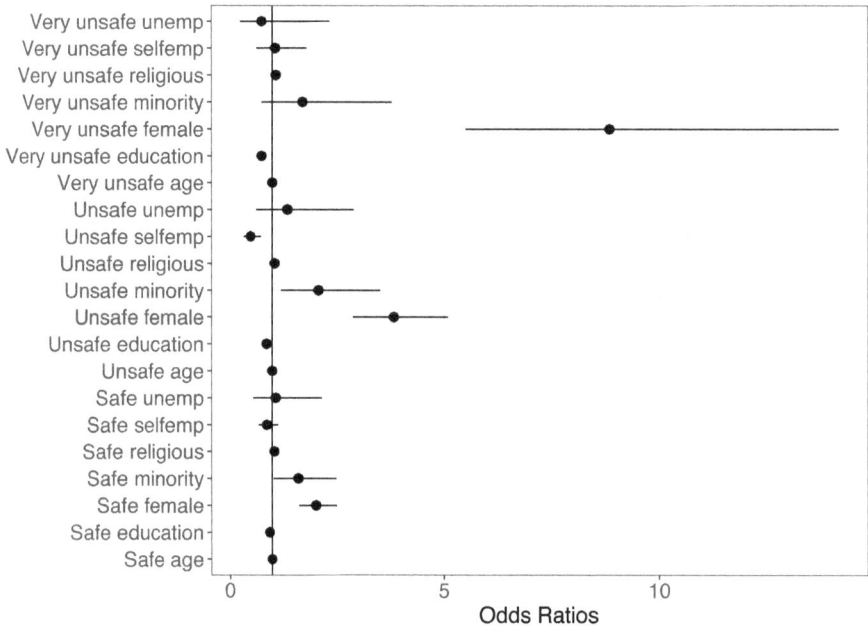

FIGURE 8.2
Odds ratio plot and confidence intervals compared to "Very safe."

odds ratios). Being a racial or ethnic minority is associated with a 2.056 and 1.604 times increase in the odds of reporting feeling "Unsafe" and "Safe" compared to "Very safe," respectively ($p<.05$). Being female is associated with a 8.829, 3.824, and 2.009 times increase in the odds of reporting feeling "Very unsafe," "Unsafe," and "Safe" compared to "Very safe," respectively ($p<.001$). Each additional year of education is associated with a 0.752, 0.867, and 0.943 times decrease in the odds of reporting feeling "Very unsafe," "Unsafe," and "Safe" compared to "Very safe," respectively ($p<.001$).

Predicted Probabilities

Predicted probabilities and changes in them have been the preferred interpretation method throughout this book. This is especially the case with the MRM because the regression coefficients become unwieldy as the number of categories increases, while the number of probabilities to interpret only increases linearly. Because the MRM is a series of pairwise comparisons, no matter the chosen interpretation method, the researcher is faced with having to interpret multiple sets of coefficients, odds ratios, or predicted probabilities. In the case of the MRM, predicted probabilities have the relative advantage of making the choice of base category irrelevant. Rather than having to swap out the base category to make a specific comparison, as one would have to do with coefficients or odds ratios, one just has to generate and interpret one set of predicted probabilities for the model.

Using education as an example, we use `margins.des` to create a design matrix for `margins.dat` to calculate the predictions. The call is similar to the call we made in Chapter 7, but here, the `data` have to be specified because `margins.des` is creating a design matrix based on a dummy model behind the scenes (the nnet model object does contain this information). As we have shown in the last chapter, there is a lot of output because we are generating predicted probabilities for each response category. Rather than showing that output here, we skip straight to plotting these probabilities:

```
> design <- margins.des(m1,ivs=expand.grid(education=10:18), data=X)
> pdat <- margins.dat(m1,design)
> ggplot(pdat,aes(x=education,y=prob,ymin=ll,ymax=ul,group=walk.alone.
dark,linetype=walk.alone.dark,color=walk.alone.dark)) +
+    theme_bw() + geom_line() + geom_point() + geom_ribbon(alpha=.1) +
+    labs(x="Education",y="Predicted Probability",linetype="",color="") +
+    scale_color_manual(values=natparks.pals("Glacier"))
```

As shown in Figure 8.3, education has a positive effect on feeling safe at night. We plot the predicted probability from 10 years of education to 18. As the level of education increases, we see that the predicted probability of

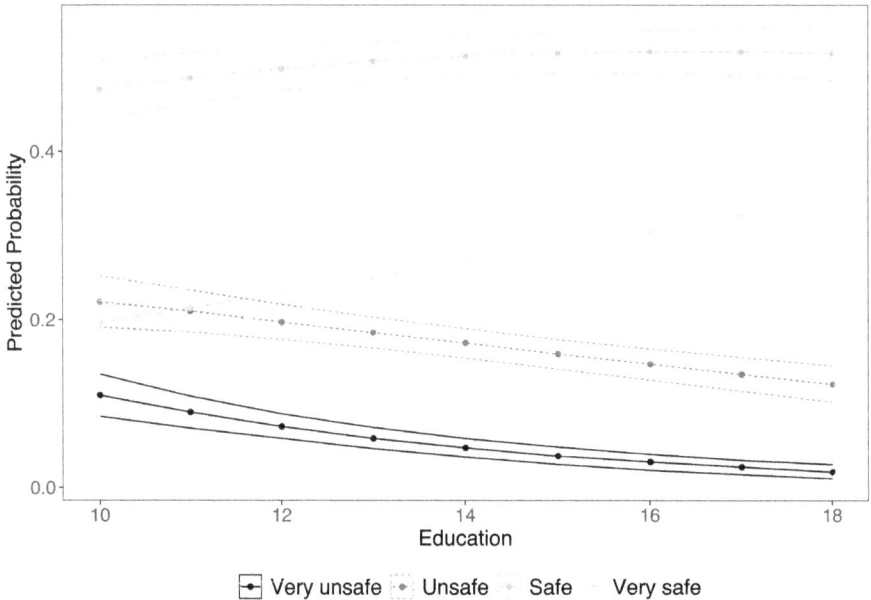

FIGURE 8.3
Predicted probability of feeling safe at night by education.

reporting "Very unsafe" and "Unsafe" each decrease. Whereas the decrease in the predicted probability of "Unsafe" is fairly linear, we see the predicted probability change is tapered toward the higher end of education for "Very unsafe." On the positive end of the response scale, we can see that the predicted probability of feeling "Very safe" increases in a linear fashion as level of education increases. The predicted probability of reporting "Safe" is fairly flat across the range of education plotted, but there is a slight positive relationship.

As we did in prior chapters, we can use `first.diff.fitted` to test for the statistical significance of changes in predicted probabilities shown in the plot. Using the following call, we can see that when comparing the predicted probability of someone with 10 years of education compared to someone with 18 years of education, the differences are statistically significant for all comparisons except for the effect of education on the "Safe" response option. This is consistent with what we observed in our plot. Note that the comparison here is 10 years of education versus 18 years. So, the differences are the opposite signs of our interpretation of increases in education, that is, 18 versus 10 years.

```
> design
  education religious    minority     female      age    selfemp        unemp
1        10  3.607542 0.07681564 0.5465549 53.12803 0.1680633 0.03258845
2        11  3.607542 0.07681564 0.5465549 53.12803 0.1680633 0.03258845
3        12  3.607542 0.07681564 0.5465549 53.12803 0.1680633 0.03258845
4        13  3.607542 0.07681564 0.5465549 53.12803 0.1680633 0.03258845
5        14  3.607542 0.07681564 0.5465549 53.12803 0.1680633 0.03258845
6        15  3.607542 0.07681564 0.5465549 53.12803 0.1680633 0.03258845
7        16  3.607542 0.07681564 0.5465549 53.12803 0.1680633 0.03258845
8        17  3.607542 0.07681564 0.5465549 53.12803 0.1680633 0.03258845
9        18  3.607542 0.07681564 0.5465549 53.12803 0.1680633 0.03258845
> first.diff.fitted(m1,design,compare=c(1,9)) # This is the change in
the probability of each response category
  first.diff std.error statistic p.value     ll     ul          dv
1      0.090     0.014     6.560   0.000  0.063  0.117 Very unsafe
2      0.098     0.020     4.846   0.000  0.058  0.138      Unsafe
3     -0.043     0.025    -1.724   0.085 -0.093  0.006        Safe
4     -0.145     0.021    -6.808   0.000 -0.187 -0.103   Very safe
```

Of course, we can also test for differences between any other values of education. By changing the `compare` option to compare the third and seventh values, corresponding to 12 versus 16 years of education:

```
> first.diff.fitted(m1,design,compare=c(3,7)) # 12 versus 16 years
of schooling
  first.diff std.error statistic p.value     ll     ul          dv
1      0.043     0.006     7.192   0.000  0.031  0.054 Very unsafe
2      0.051     0.010     4.934   0.000  0.030  0.071      Unsafe
3     -0.020     0.013    -1.614   0.106 -0.045  0.004        Safe
4     -0.073     0.011    -6.700   0.000 -0.094 -0.052   Very safe
```

We can also examine the effect of categorical variables like respondent sex by changing the `margins.des` call and the plot type:

```
> design <- margins.des(m1,ivs=expand.grid(female=c(0,1)),
data=X)
> pdat <- margins.dat(m1,design)
> pdat <- mutate(pdat,sex=rep(c("Male","Female"),each=4),
+               xaxs=c(0,.15,.3,.45,.05,.2,.35,.5))
> ggplot(pdat,aes(x=xaxs,y=prob,ymin=ll,ymax=ul,fill=sex)) +
+    theme_bw() + geom_col() +
+    scale_fill_manual(values=c("grey65","grey35")) +
+    geom_errorbar(width=.02) + theme(legend.position="bottom")
+
+    labs(x="",y="Predicted Probability",fill="") +
+    scale_x_continuous(breaks=c(.025,.175,.325,.525),
+                    labels=c("Very Unsafe","Unsafe",
"Safe","Very Safe"))
```

In this Figure, we can see that male respondents have high predicted probabilities of reporting feeling "Safe" and "Very safe" at night. Although female respondents also have a high predicted probability of reporting "Safe," they have significantly lower predicted probabilities of reporting "Very safe" compared to male responses. Likewise, they have significantly higher predicted probabilities of reporting both "Very unsafe" and "Unsafe" (Figure 8.4). We confirm that these differences are statistically significant by calling `first.diff.fitted`:

```
> design <- margins.des(m1,ivs=expand.grid(female=c(0,1)), data=X)
> first.diff.fitted(m1,design,compare=c(1,2)) # Significance of
those differences
  first.diff std.error statistic p.value      ll      ul         dv
1     -0.068     0.009    -7.261   0.000  -0.086  -0.050 Very unsafe
2     -0.107     0.016    -6.587   0.000  -0.139  -0.075      Unsafe
3     -0.011     0.022    -0.494   0.622  -0.055   0.033        Safe
4      0.186     0.020     9.285   0.000   0.147   0.225   Very safe
```

Marginal Effects

In addition to comparing marginal effects at the mean as we have done up to this point, we can also use average marginal effects (AME and conditional marginal effects). We walk through these alternatives below. Recall from the previous chapters that the `marginaleffects` function can generate the average marginal effect of some or all variables in a model. Below we compute the AME of our independent variables and plot them:

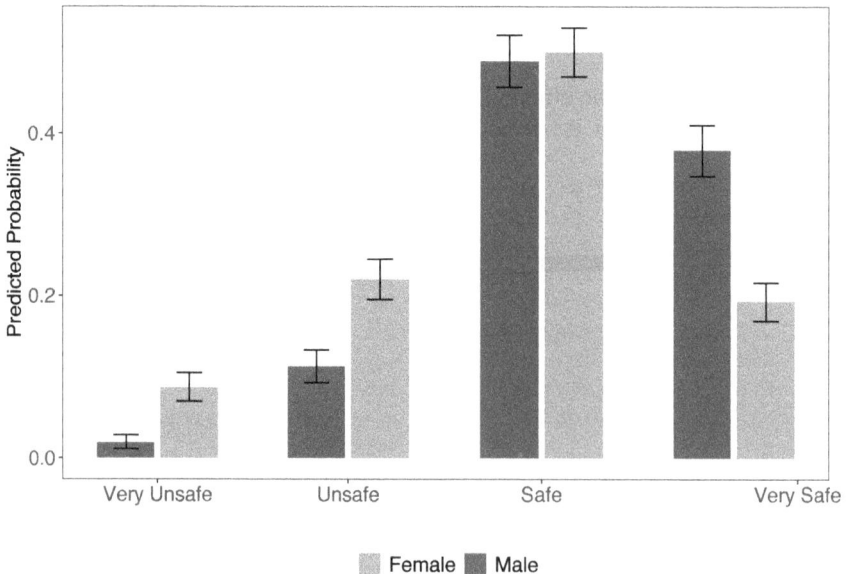

FIGURE 8.4
Predicted probability of feeling safe at night by respondent sex.

```
summary(marginaleffects(m1))
```

	Group	Term	Effect	Std. Error	z value	Pr(>|z|)	2.5 %	97.5 %
1	Very safe	education	0.0172228	0.0025253	6.81998	9.1050e-12	0.0122732	0.0221724
2	Very safe	religious	-0.0098520	0.0032706	-3.01227	0.00259300	-0.0162623	-0.0034417
3	Very safe	minority	-0.0987512	0.0397876	-2.48196	0.01306616	-0.1767334	-0.0207690
4	Very safe	female	-0.1752572	0.0176377	-9.93651	< 2.22e-16	-0.2098264	-0.1406879
5	Very safe	age	0.0001813	0.0005576	0.32507	0.74512697	-0.0009117	0.0012742
6	Very safe	selfemp	0.0461870	0.0239815	1.92594	0.05411140	-0.0008159	0.0931898
7	Very safe	unemp	-0.0188299	0.0617044	-0.30516	0.76024143	-0.1397683	0.1021084
8	Very unsafe	education	-0.0131057	0.0020242	-6.47447	9.5144e-11	-0.0170731	-0.0091383
9	Very unsafe	religious	0.0022590	0.0018485	1.22209	0.22167317	-0.0013639	0.0058819
10	Very unsafe	minority	0.0054182	0.0219754	0.24656	0.80525254	-0.0376529	0.0484892
11	Very unsafe	female	0.0880451	0.0140843	6.25131	4.0703e-10	0.0604404	0.1156497
12	Very unsafe	age	0.0001447	0.0003191	0.45351	0.65018314	-0.0004807	0.0007701
13	Very unsafe	selfemp	0.0194366	0.0146085	1.33050	0.18335523	-0.0091956	0.0480687
14	Very unsafe	unemp	-0.0240925	0.0299956	-0.80320	0.42185872	-0.0828827	0.0346978
15	Unsafe	education	-0.0104665	0.0024292	-4.30860	1.6429e-05	-0.0152276	-0.0057053
16	Unsafe	religious	0.0021715	0.0027793	0.78133	0.43460671	-0.0032757	0.0076188
17	Unsafe	minority	0.0525739	0.0293408	1.79183	0.07315972	-0.0049331	0.1100809
18	Unsafe	female	0.0942544	0.0166874	5.64825	1.6209e-08	0.0615478	0.1269611
19	Unsafe	age	-0.0001800	0.0004792	-0.37569	0.70715071	-0.0011193	0.0007592
20	Unsafe	selfemp	-0.0901712	0.0267947	-3.36526	0.00076472	-0.1426879	-0.0376545
21	Unsafe	unemp	0.0389338	0.0398368	0.97733	0.32840475	-0.0391449	0.1170125
22	Safe	education	0.0063494	0.0030673	2.07003	0.03844990	0.0003376	0.0123612
23	Safe	religious	0.0054214	0.0037322	1.45263	0.14632661	-0.0018935	0.0127363
24	Safe	minority	0.0407592	0.0424228	0.96079	0.33666020	-0.0423879	0.1239063
25	Safe	female	-0.0070423	0.0218620	-0.32212	0.74735800	-0.0498911	0.0358065
26	Safe	age	-0.0001459	0.0006410	-0.22767	0.81990442	-0.0014024	0.0011105
27	Safe	selfemp	0.0245477	0.0298843	0.82142	0.41140583	-0.0340246	0.0831199
28	Safe	unemp	0.0039886	0.0648160	0.06154	0.95093100	-0.1230484	0.1310257

Picking out the `female` example to compare with our MEM discussion, we can see that the AME of female is similar to the results from our `first.diff.fit-ted` call, with the exception of the sign of the effect because we compared "male" to "female" in the `first.diff.fitted` call and "female" to "male" here. To examine the average marginal effect conditional on another variable, we can use `newdata=datagrid` option to predict these marginal effects for each level of a conditioning variable. In our next example, we examine the marginal effects based on racial or ethnic minority status:

```
> summary(marginaleffects(m1,newdata=datagrid(minority=1)))
            Group       Term       Effect Std. Error z value   Pr(>|z|)         2.5 %       97.5 %
1      Very safe education  1.440e-02  0.0027658  5.2053 1.9373e-07  0.0089759  0.0198178
2      Very safe religious -7.970e-03  0.0030054 -2.6520 0.00800231 -0.0138609 -0.0020798
3      Very safe  minority -8.196e-02  0.0221983 -3.6923 0.00022222 -0.1254710 -0.0384554
4      Very safe    female -1.456e-01  0.0238181 -6.1136 9.7432e-10 -0.1922963 -0.0989309
5      Very safe       age  1.560e-04  0.0004552  0.3426 0.73188523 -0.0007362  0.0010481
6      Very safe    selfemp  4.276e-02  0.0207138  2.0645 0.03896674  0.0021661  0.0833628
7      Very safe     unemp -1.756e-02  0.0499094 -0.3519 0.72489817 -0.1153847  0.0802565
8    Very unsafe education -1.052e-02  0.0035221 -2.9867 0.00281993 -0.0174229 -0.0036164
9    Very unsafe religious  1.759e-03  0.0014589  1.2057 0.22791889 -0.0011003  0.0046184
10   Very unsafe  minority  4.105e-03  0.0188659  0.2176 0.82775998 -0.0328718  0.0410813
11   Very unsafe    female  7.043e-02  0.0240785  2.9252 0.00344254  0.0232411  0.1176272
12   Very unsafe       age  1.160e-04  0.0002638  0.4399 0.66000922 -0.0004010  0.0006331
13   Very unsafe    selfemp  1.504e-02  0.0127250  1.1822 0.23712354 -0.0098970  0.0399842
14   Very unsafe     unemp -1.913e-02  0.0253463 -0.7546 0.45051502 -0.0688030  0.0305526
15       Unsafe education -1.443e-02  0.0036787 -3.9226 8.7590e-05 -0.0216405 -0.0072201
16       Unsafe religious  2.494e-03  0.0034687  0.7189 0.47223079 -0.0043051  0.0092921
17       Unsafe  minority  6.198e-02  0.0435629  1.4228 0.15478626 -0.0233993  0.1473643
18       Unsafe    female  1.238e-01  0.0265990  4.6536 3.2623e-06  0.0716474  0.1759136
19       Unsafe       age -1.968e-04  0.0006000 -0.3280 0.74293936 -0.0013728  0.0009792
20       Unsafe    selfemp -1.101e-01  0.0353112 -3.1175 0.00182384 -0.1792918 -0.0408744
21       Unsafe     unemp  4.509e-02  0.0498994  0.9037 0.36615740 -0.0527072  0.1428948
22         Safe education  1.055e-02  0.0042579  2.4785 0.01319534  0.0022077  0.0188985
23         Safe religious  3.718e-03  0.0039179  0.9489 0.34265424 -0.0039611  0.0113967
24         Safe  minority  1.588e-02  0.0441126  0.3599 0.71892571 -0.0705833  0.1023351
25         Safe    female -4.860e-02  0.0324706 -1.4968 0.13445238 -0.1122423  0.0150401
26         Safe       age -7.521e-05  0.0006601 -0.1139 0.90928428 -0.0013690  0.0012186
27         Safe    selfemp  5.228e-02  0.0340945  1.5332 0.12521662 -0.0145489  0.1190991
28         Safe     unemp -8.404e-03  0.0632471 -0.1329 0.89428544 -0.1323665  0.1155575

Model type:  multinom
Prediction type:  probs

> summary(marginaleffects(m1,newdata=datagrid(minority=0)))
            Group       Term       Effect Std. Error z value   Pr(>|z|)         2.5 %       97.5 %
1      Very safe education  1.852e-02  0.0028472  6.50304 7.8714e-11  0.0129349  0.0240956
2      Very safe religious -1.062e-02  0.0035384 -3.00224 0.00267999 -0.0175582 -0.0036880
3      Very safe  minority -1.073e-01  0.0438614 -2.44689 0.01440950 -0.1932908 -0.0213572
4      Very safe    female -1.890e-01  0.0203757 -9.27767 < 2.22e-16 -0.2289746 -0.1491033
5      Very safe       age  2.013e-04  0.0006029  0.33385 0.73849290 -0.0009804  0.0013830
6      Very safe    selfemp  5.172e-02  0.0261026  1.98143 0.04754297  0.0005603  0.1028807
7      Very safe     unemp -2.142e-02  0.0666699 -0.32131 0.74797571 -0.1520923  0.1092489
8    Very unsafe education -9.954e-03  0.0013133 -7.57909 3.4798e-14 -0.0125279 -0.0073797
9    Very unsafe religious  1.809e-03  0.0013522  1.33788 0.18093632 -0.0008412  0.0044595
10   Very unsafe  minority  6.388e-03  0.0158133  0.40399 0.68621835 -0.0246051  0.0373820
11   Very unsafe    female  6.818e-02  0.0089512  7.61655 2.6055e-14  0.0506333  0.0857214
12   Very unsafe       age  9.832e-05  0.0002310  0.42563 0.67037826 -0.0003544  0.0005511
13   Very unsafe    selfemp  1.137e-02  0.0105522  1.07731 0.28133977 -0.0093139  0.0320499
14   Very unsafe     unemp -1.630e-02  0.0216877 -0.75138 0.45242547 -0.0588028  0.0262115
15       Unsafe education -1.255e-02  0.0024565 -5.10871 3.2437e-07 -0.0173642 -0.0077349
```

16	Unsafe	religious	2.683e-03	0.0027844	0.96347	0.33531097	-0.0027747	0.0081401
17	Unsafe	minority	5.547e-02	0.0287801	1.92750	0.05391790	-0.0009344	0.1118816
18	Unsafe	female	1.096e-01	0.0168626	6.49801	8.1387e-11	0.0765234	0.1426237
19	Unsafe	age	-1.653e-04	0.0004779	-0.34590	0.72941620	-0.0011021	0.0007714
20	Unsafe	selfemp	-8.870e-02	0.0263493	-3.36625	0.00076197	-0.1403418	-0.0370547
21	Unsafe	unemp	3.624e-02	0.0400265	0.90533	0.36529310	-0.0422135	0.1146875
22	Safe	education	3.988e-03	0.0031488	1.26656	0.20531434	-0.0021834	0.0101596
23	Safe	religious	6.131e-03	0.0037955	1.61542	0.10622079	-0.0013077	0.0135702
24	Safe	minority	4.546e-02	0.0433139	1.04959	0.29390486	-0.0394317	0.1303556
25	Safe	female	1.129e-02	0.0224420	0.50299	0.61497351	-0.0326976	0.0552737
26	Safe	age	-1.343e-04	0.0006491	-0.20688	0.83610401	-0.0014065	0.0011379
27	Safe	selfemp	2.561e-02	0.0301572	0.84921	0.39576600	-0.0334973	0.0847168
28	Safe	unemp	1.480e-03	0.0661716	0.02237	0.98215193	-0.1282137	0.1311744

When having this much output, it helps to visually plot the marginal effects. To make the example clearer, we again focus on the AME of female, which can be visually illustrated in Figure 8.5. As shown in the figure, the effect of being female appears to be different for racial and ethnic minorities than nonminorities. The AME of female is similar for "Very unsafe" and "Unsafe" – female respondents are significantly more likely to select "Very unsafe" and "Unsafe" – but different for "Safe" and "Very safe." Here, we see that the difference between male and female respondents is smaller for nonminority respondents selecting "Safe" than for minority respondents. The difference in AME of female is slightly larger for nonminority respondents than minority respondents selecting "Very safe."

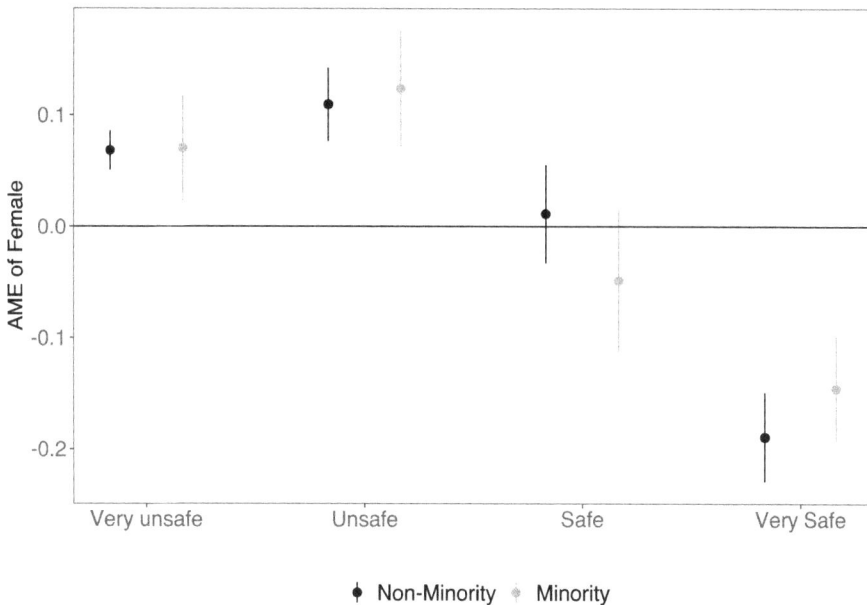

FIGURE 8.5
AME of female by minority status.

Combining Categories

As we have seen, the presentation and interpretation of results from the MRM can be unwieldy. Because the MRM is a series of BRM, one option for simplifying the presentation of results is to combine categories of the outcome that are not significantly different from one another to reduce the number of pairwise comparisons. Doing so requires performing a likelihood ratio test, as we have done in the past to see if the more complicated model with more categories significantly improves the fit of the model compared to one with a reduced number of categories. Suppose we have theoretical reasons to expect that "Very unsafe" and "Unsafe" are similar to one another. In that case, we can fit a model that recodes the four-category outcome to a three-category outcome of "Unsafe," "Safe," and "Very safe." Then, we can perform the LR test as we have done in the past:

```
X <- mutate(X,walk2=recode_factor(walk.alone.dark,
"Very unsafe"="Unsafe"))
m2<-multinom(walk2 ~ education + religious + minority  +
female + age + selfemp + unemp,data=X)
lr.test(m1,m2)
# weights:   27 (16 variable)
initial  value 2359.819196
iter  10 value 2145.324115
iter  20 value 2125.539520
final  value 2125.536539
converged
> lr.test(m1,m2)
    LL.Full LL.Reduced G2.LR.Statistic DF p.value
1 -2417.426   -2125.537         583.7788  8       0
```

Here, we see that the full model significantly improves the fit of the model compared to the reduced model with fewer categories ($\chi^2 = 583.8$, $p < .001$). In this case, the empirical evidence suggests that we lose information by combining categories. If the test was nonsignificant, we could combine outcome categories and retain the simpler model.

The Independence of Irrelevant Alternatives

We end this chapter by discussing the independence of irrelevant alternatives' assumption of the MRM. As we discussed earlier, implicit in the way the MRM is set up is that the choice between every pairwise BRM is unaffected by the other options. From Eq. 8.3, we see that the equation:

$$\ln \frac{\Pr(y = m \,|\, x)}{\Pr(y = b \,|\, x)} = \mathbf{X}\mathbf{b}_{m|b} + \varepsilon_{m|b} \text{ for } m=1 \text{ to } J, \text{ only considers categories } m \text{ and}$$

b, not $m - 1$, $m+1$, etc. This may be a reasonable assumption, but it is never-theless a strong assumption (Ray 1973). Put more substantively, the model assumes that the choice between any two options, say whether one drives to work or takes the bus, is independent of any other option, including if good public transit is an option (Buehler and Hamre 2015). Another example would be an example where whether one votes for a Democrat or Republican candidate does not depend on if a viable third-party candidate exists (Dow and Endersby 2004). As with most assumptions, we can test if this is the case for our model. Mathematically, the model implies that how we set up the BRM should not affect the coefficients. To perform the test, we use the `mlogit` function from that same package (Croissant 2020) to fit the differ-ent alternative specifications for the comparison groups. Below we fit our baseline model first, and then fit four additional models, each altering the response categories:

```
X2<-mlogit.data(X,choice="walk.alone.dark",shape="wide")

ml1 <- mlogit(walk.alone.dark ~ 0|education + religious +
minority + female + age + selfemp + unemp, data=X2, reflevel
= "Very safe")

ml2 <- mlogit(walk.alone.dark ~ 0|education + religious +
minority + female + age + selfemp + unemp, data=X2,
            reflevel = "Very safe",
            alt.subset=c("Safe","Unsafe","Very safe"))

ml3 <- mlogit(walk.alone.dark ~ 0|education + religious +
minority + female + age + selfemp + unemp, data=X2,
            reflevel = "Very safe",
            alt.subset=c("Safe","Very unsafe","Very safe"))

ml4 <- mlogit(walk.alone.dark ~ 0|education + religious +
minority + female + age + selfemp + unemp, data=X2,
            reflevel = "Very safe",
            alt.subset=c("Unsafe","Very unsafe",
"Very safe"))

ml5 <- mlogit(walk.alone.dark ~ 0|education + religious +
minority + female + age + selfemp + unemp, data=X2,
            reflevel = "Safe",
            alt.subset=c("Unsafe","Very unsafe","Safe"))
```

First, recall that we need to reshape the data to use the `mlogit` function. Then, the `ml1` object stores our baseline MRM model. Each of the `ml2`, `ml3`, `ml4`, and `ml5` objects are fitting alternative models where one of the alternative categories is excluded: "Very unsafe," "Unsafe," "Safe," or "Very safe," respectively. Then, we use `hmftest` to perform a Hausman and McFadden (1984) test comparing each of these alternatives to our baseline MRM model:

```
> hmftest(ml1,ml2) # Test for Very unsafe

        Hausman-McFadden test

data:  X2
chisq = -2.1267, df = 16, p-value = 1
alternative hypothesis: IIA is rejected

> hmftest(ml1,ml3) # Test for Unsafe

        Hausman-McFadden test

data:  X2
chisq = -0.82502, df = 16, p-value = 1
alternative hypothesis: IIA is rejected

> hmftest(ml1,ml4) # Test for Safe

        Hausman-McFadden test

data:  X2
chisq = 13.796, df = 16, p-value = 0.6139
alternative hypothesis: IIA is rejected

> hmftest(ml1,ml5) # Test for Very safe

        Hausman-McFadden test

data:  X2
chisq = -421.09, df = 16, p-value = 1
alternative hypothesis: IIA is rejected
```

As we can see, there is not enough evidence to reject the null of IIA. Note, however, that Fry and Harris (1996, 1998) and Cheng and Long (2007) all find that tests of the IIA perform poorly, even with large samples and that different versions of the test lead to drastically different conclusions, which should all be asymptotically equivalent. This has led scholars to conclude that "these tests are not useful for assessing violations of the IIA property" (Long and Freese 2006:408). Alternative models like the nested logit model can be used, but they require strong theory about the nature of the nesting of choices (Hausman and McFadden 1984).

9

Regression for Count Outcomes

Although linear regression models are often used to model count variables, they can lead to biased results, especially if the count has a low mean (Coxe, West, and Aiken 2009). Linear regressions also do not easily allow for the modeling of empirical realities of counts, such as overdispersion and problems with modeling zeros. Moreover, linear models do not allow mixture processes leading to zeros and truncation of zeros in the data. We walk through models designed for each of these scenarios in this chapter. Unlike the BRM, ORM, and MRM, count models are not categorical models per se. However, as we show, the same techniques of interpretation and diagnostics (albeit with slightly different names) can be applied to count models. Because count data are different from the other outcomes we have already examined, we change the example used in this chapter to modeling the number – or count – of children per respondent using the ESS. Figure 9.1 illustrates the empirical distribution of this variable.

We begin with the simplest count model, the Poisson regression model (PRM). Then, we introduce complications to the model. First, we examine how overdispersion, when additional sources of heterogeneity lead to more variation in the data than expected, can be modeled using a mixture of a Poisson and gamma distribution leading to the negative binomial regression model (NBRM). Then, we examine a class of models that consider the unique nature of zeros in counts. In the real world, having a count can mean multiple things. In the case of number of children, having zero children may reflect an inability to have children (Greil, Slauson-Blevins, and McQuillan 2010), a choice to remain childless (Moore 2017), or someone who is yet to have children among other reasons. Regardless, the count of children for someone who can and is willing to have children reflects a different process than those who either cannot or do not want children. Zero-inflated count models account for these separate processes by modeling zero versus positive responses initially, and then for those with positive responses the count is generated. After we determine if a case is always going to be a zero, the presence of overdispersion determines whether we fit a zero-inflated Poisson (ZIP) regression or zero-inflated negative binomial (ZINB) regression. Using the countfit function included in the catregs package, we provide a tool for comparing these common count scenarios.

We end with two sets of alternative models to account for other ways of thinking about zeros. Zero-truncated models, truncated Poisson (ZTP), or truncated negative binomial (ZTNB) account for scenarios where zero

DOI: 10.1201/9781003029847-9

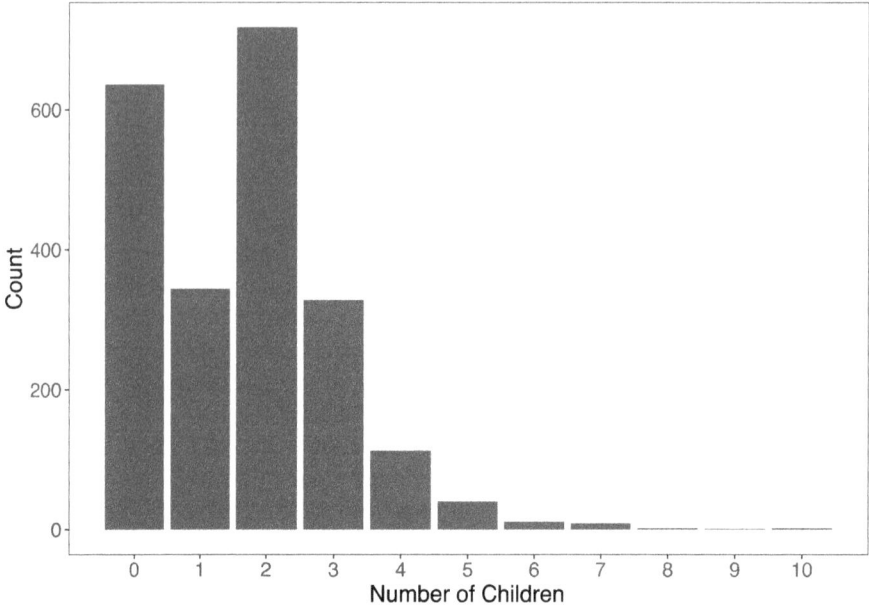

FIGURE 9.1
Histogram of the number of children. Source is the ESS.

reflects missing data. Perhaps, we sampled from parents' groups so no one in the data who are childless would be included, yet we know that there are childless people in the world. Finally, the hurdle regression model (HRM) is related to zero-inflated models, but allows for zero deflation as well as zero inflation. Whereas the zero-inflated models combine a BRM with a PRM or NBRM, the HRM combines a BRM with a truncated model. It is fit on two separate parts of the data and is a two-component mixture model.

Poisson Regression

We briefly touched on the Poisson regression (PRM) in Chapter 4 when discussing the log-linear model. There, we used the log-linear model to model the counts of respondents in cells of a contingency table. Here, we use PRM to model the count of an outcome of interest given a set of covariates. The PRM can be thought of as a generalization of a linear model using a log link to force the outcome to be zero or positive only, which is required for counts. Therefore:

$$\eta(\mathbf{y}) = \mathbf{Xb} \tag{9.1}$$

becomes:

$$\log\big(\mathrm{E}(y)\big) = \mathbf{Xb} \tag{9.2}$$

To get the expected value for *y*, then, we take the exponential of each side, resulting in the PRM:

$$\mathrm{E}(y) = \exp(\mathbf{Xb}) \tag{9.3}$$

To use the log link, the PRM assumes that the outcome variable follows a Poisson distribution.

Estimation

Because a PRM is a GLM with no added complications, we can use R's built-in glm function to estimate a PRM:

```
m1 <- glm(num.children ~ religious + minority  + female + age +
education + divorced + married + widow ,data=X,family="poisson")
summary(m1)

Call:
glm(formula = num.children ~ religious + minority + female +
    age + education + divorced + married + widow, family =
"poisson",
    data = X)

Deviance Residuals:
    Min      1Q   Median      3Q      Max
-2.8612  -1.3162  -0.0547   0.5767   4.6290

Coefficients:
            Estimate Std. Error z value Pr(>|z|)
(Intercept) -0.414943   0.103038  -4.027 5.65e-05 ***
religious    0.010146   0.005832   1.740 0.081907 .
minority     0.179521   0.064919   2.765 0.005687 **
female       0.202184   0.035174   5.748 9.03e-09 ***
age          0.016332   0.001126  14.502  < 2e-16 ***
education   -0.015956   0.004839  -3.297 0.000976 ***
divorced     0.172232   0.049283   3.495 0.000475 ***
married      0.285771   0.060374   4.733 2.21e-06 ***
widow       -0.062473   0.058664  -1.065 0.286909
---
Signif. codes:  0 '***' 0.001 '**' 0.01 '*' 0.05 '.' 0.1 ' ' 1

(Dispersion parameter for poisson family taken to be 1)
```

```
   Null deviance: 3128.5  on 2156  degrees of freedom
Residual deviance: 2699.6  on 2148  degrees of freedom
  (47 observations deleted due to missingness)
AIC: 6780.2

Number of Fisher Scoring iterations: 5
```

In this example, we are predicting the respondent's number of children in the ESS. We use similar independent variables in our previous examples – religiosity, minority status, respondent sex, age, and education. In addition, we add marital status to the equation because it probably has some bearing on childbearing. Much of the output is the same as our output for the BRM because it is using the same function as the BRM, although with a different link function. The only difference in the output is the note that the "Dispersion parameter for [the] poisson family [is] taken to be 1," which indicates that the major assumption of the PRM: equidispersion, or that the mean and variance is the same. Whether this is a reasonable assumption for the data is easily testable and discussed in greater detail when we discuss the NBRM. For now, we move on to interpreting the PRM.

As with our previous discussions, it is possible to interpret the raw regression coefficients here. They are referred to as log-counts, because the PRM is a log of the expected value of the outcome. However, people tend to not think of counts in logarithm terms and other interpretation methods are more useful. Nevertheless, we can interpret the effect of the `married` variable as:

> Being married, compared to being single, increases the log-count of the number of children under 18 living with the respondent by .29, all else equal.

Incidence Rate Ratios

The incidence rate ratio (IRR) for the PRM is analogous to the odds ratio for the BRM. As with the BRM, we arrive at the IRR by exponentiating the regression coefficients. Unlike the BRM, they are much more useful and natural to think about here than they are with odds. Whereas factor changes in odds are difficult to understand without knowing the baseline probabilities (e.g., is doubling the odds a large change? It depends on how big the starting point is), factor changes in counts are intuitively useful. While it is still true that doubling the number of children depends on how many children we started out with, people more intuitively understand the effect of having twice the number of children regardless of if it is going from having one child to two or going from having five to ten children. We can call the `list.coef` function to calculate the IRR:

```
list.coef(m1)
$out
```

	variables	b	SE	z	ll	ul	p.val	exp.b	ll.exp.b	ul.exp.b	percent	CI
1	(Intercept)	-0.415	0.103	-4.027	-0.617	-0.213	0.000	0.660	0.540	0.808	-33.962	95%
2	religious	0.010	0.006	1.740	-0.001	0.022	0.088	1.010	0.999	1.022	1.020	95%
3	minority	0.180	0.065	2.765	0.052	0.307	0.009	1.197	1.054	1.359	19.664	95%
4	female	0.202	0.035	5.748	0.133	0.271	0.000	1.224	1.143	1.311	22.407	95%
5	age	0.016	0.001	14.502	0.014	0.019	0.000	1.016	1.014	1.019	1.647	95%
6	education	-0.016	0.005	-3.297	-0.025	-0.006	0.002	0.984	0.975	0.994	-1.583	95%
7	divorced	0.172	0.049	3.495	0.076	0.269	0.001	1.188	1.079	1.308	18.795	95%
8	married	0.286	0.060	4.733	0.167	0.404	0.000	1.331	1.182	1.498	33.079	95%
9	widow	-0.062	0.059	-1.065	-0.177	0.053	0.226	0.939	0.837	1.054	-6.056	95%

Interpreting the `exp.b` column, we can interpret the effect of being married along these lines:

> Being married, compared to being single, is associated with an increase in the number of children under 18 living with the respondent by a factor of 1.33, all else equal.

Alternatively, we can interpret the percent change:

> Being married, compared to being single, is associated with an increase in the number of children under 18 living with the respondent by 33 percent, all else equal.

Finally, we can call the `ggplot` function or a similar graphing function to create a visualization of the IRRs like the one shown in Figure 9.2. Here, we sorted the IRRs to show the relative magnitude of the effects. As shown in the figure, being married, female, and a racial or ethnic minority each have relatively large effects on the IRR of the number of children. Being older and more religious have smaller but still positive per unit effects, in part because of the larger range for those variables. More education is associated with a small but significant negative effect on the number of children. Being widowed does not have a significant effect.

```
pdat <- list.coef(m1)$out
ggplot(pdat[2:nrow(pdat),],aes(y=reorder(variables,exp.b),x=ex
p.b,xmin=ll.exp.b,xmax=ul.exp.b)) +
   theme_bw() + geom_pointrange() + labs(x="Incident Rate
Ratio",y="") + geom_vline(xintercept=1)
```

Marginal Effects

For the previous chapters, we discussed marginal effects with predicted probabilities because probabilities were the natural metric for interpreting binary, ordinal, or nominal categories. However, for count models, counts are the natural metric, not the probability of each outcome category. Therefore, it makes more sense to calculate predicted counts and get marginal effects from those counts than from the predicted probability of the counts.

To calculate the marginal effect at the mean for a continuous variable like education, for example, we use `margins.des` and `margins.dat` and plot the resulting predictions:

```
design <- margins.des(m1,ivs=expand.grid(education=10:18))
pdat <- margins.dat(m1,design)
ggplot(pdat,aes(x=education,y=fitted,ymin=ll,ymax=ul)) +
   theme_bw() + geom_point() + geom_line() + geom_ribbon(alpha=.1) +
   labs(x="Education",y="Predicted Count of Children")
```

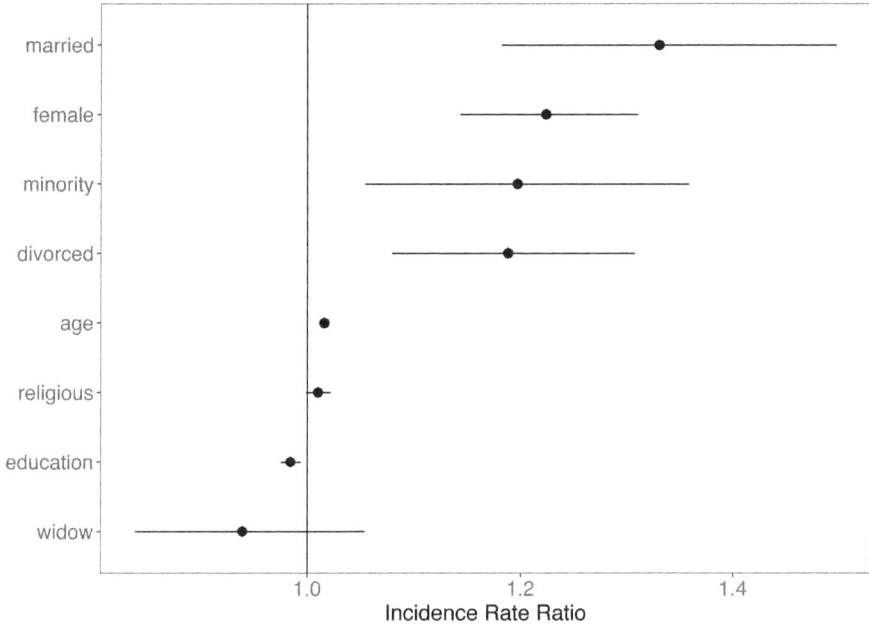

FIGURE 9.2
Sorted plot of incidence rate ratios.

As shown in Figure 9.3, years of education has a negative effect on the predicted number of children. Whereas someone with only 10 years of education is expected to have 1.6 children, holding other variables at their means, someone with 18 years of education is expected to have only 1.4 children. We can use first.diff.fitted to test that this difference is statistically significant and see that the difference of 0.196 is significant ($p < .001$).

```
first.diff.fitted(m1,pdat,compare=c(1,9))
  first.diff std.error statistic p.value    ll     ul
1      0.196      0.06     3.282   0.001 0.079 0.312
```

We can similarly examine the marginal effects of categorical variables like marital status. Below is the code to generate Figure 9.4, illustrating how the predicted count of children varies with marital status:

```
d1 <- margins.des(m1,ivs=expand.grid(divorced=0,married=0,widow=0))
d2 <- margins.des(m1,ivs=expand.grid(divorced=1,married=0,widow=0))
d3 <- margins.des(m1,ivs=expand.grid(divorced=0,married=1,widow=0))
d4 <- margins.des(m1,ivs=expand.grid(divorced=0,married=0,widow=1))
design <- rbind(d1,d2,d3,d4)
pdat<- margins.dat(m1,design)
```

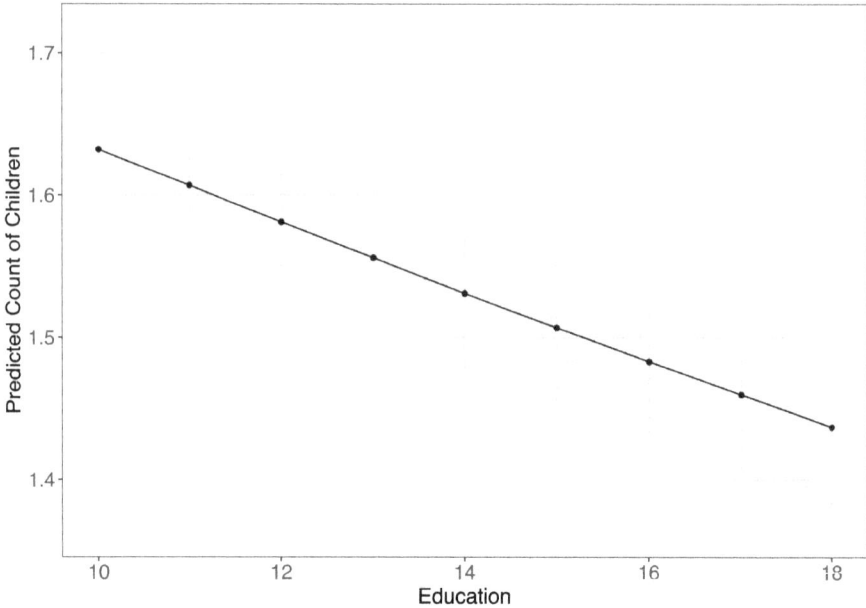

FIGURE 9.3
Predicted number of children by years of education.

```
pdat <- mutate(pdat,marital=c("Single","Divorced","Married","Widow"))
ggplot(pdat,aes(x=marital,y=fitted,ymin=ll,ymax=ul)) +
   theme_bw() + geom_pointrange() + labs(x="",y="Predicted Count of
Children")
```

Because we have a multi-category variable of interest, we generate a separate
design matrix for each combination of the three indicator variables in the
model using `margins.des` and combine them as input for `margins.dat`.
As we can see in the resulting plot, married and divorced respondents have
significantly more children than single and widowed respondents (nonoverlapping confidence intervals). We can use `first.diff.fitted` to confirm that
the difference between being divorced and single (0.277, $p<.001$) and between
married and single (0.488, $p<.001$) are significant but the difference between
being widowed and single (–0.089, ns) is not. We can do the same and compare
against widowed respondents by changing the `compare` option.

```
first.diff.fitted(m1,pdat,compare=c(1,2))
   first.diff std.error statistic p.value      ll      ul
1      -0.277     0.084    -3.301    0.001 -0.442 -0.113
> first.diff.fitted(m1,pdat,compare=c(1,3))
   first.diff std.error statistic p.value      ll      ul
1      -0.488     0.115    -4.232        0 -0.714 -0.262
```

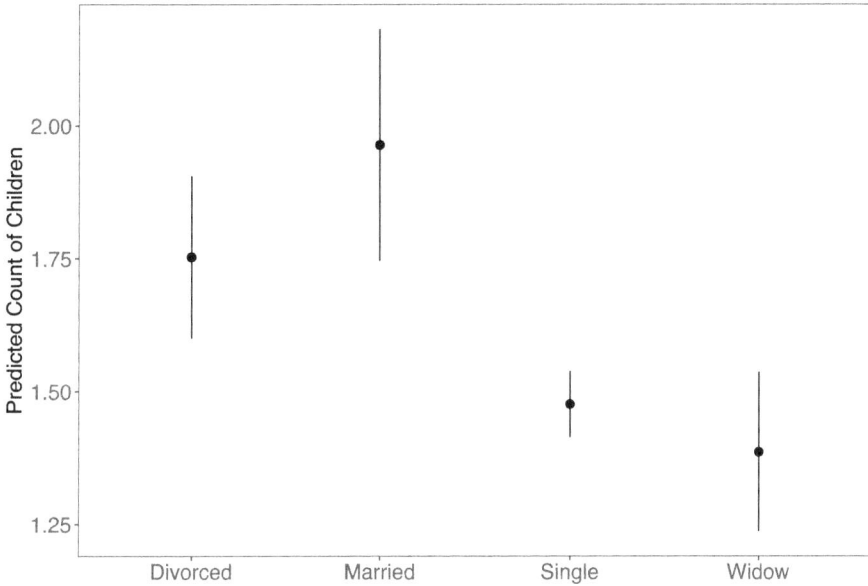

FIGURE 9.4
Predicted number of children by marital status.

```
> first.diff.fitted(m1,pdat,compare=c(1,4))
  first.diff std.error statistic p.value      ll    ul
1      0.089     0.082     1.089   0.276 -0.071  0.25
```

Finally, we can calculate and interpret the AME using the `marginalef-fects` function:

```
> summary(marginaleffects(m1))
        Term    Effect Std. Error z value    Pr(>|z|)      2.5 %     97.5 %
1 religious   0.01648   0.009481   1.738  0.08213744  -0.00210    0.03506
2  minority   0.29163   0.105587   2.762  0.00574480   0.08468    0.49858
3    female   0.32845   0.057422   5.720  1.0660e-08   0.21590    0.44099
4       age   0.02653   0.001900  13.966  < 2.22e-16   0.02281    0.03026
5 education  -0.02592   0.007881  -3.289  0.00100631  -0.04137   -0.01047
6  divorced   0.27979   0.080192   3.489  0.00048479   0.12262    0.43696
7   married   0.46423   0.098384   4.719  2.3749e-06   0.27140    0.65706
8     widow  -0.10149   0.095322  -1.065  0.28703059  -0.28831    0.08534
```

Doing so paints a similar story both in terms of statistical significance as well as the magnitude of the AME compared to the magnitude of the MEM. For example, the AME of married is 0.46 children compared to the MEM of married being 0.48 children.

Predicted Probabilities

The Poisson distribution is defined as:

$$\Pr(y|\mu) = \frac{e^{-\mu}\mu^{y}}{y!} \text{ For } y=0, 1, 2... \tag{9.4}$$

Accordingly, given that we are interested in predicting each value of y, we can calculate the predicted probability of each value by:

$$\widehat{\Pr}(y = k|\boldsymbol{x}) = \frac{e^{-x\hat{B}}x\hat{B}^{k}}{k!} \tag{9.5}$$

We can use the dpois function and use the exponents of the model coefficients as input to generate predicted probabilities. We can use the following call, for example, to calculate the predicted probability of having no children at home, which is ~0.2:

```
dpois(0,mean(m1$fitted.values))
[1] 0.1970144
```

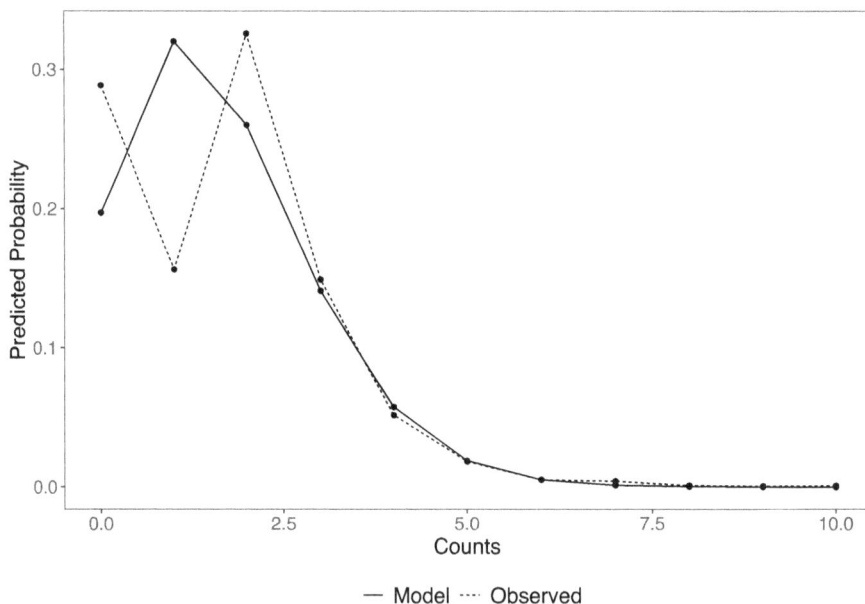

FIGURE 9.5
Predicted probabilities compared to observed proportions.

It is a common practice for researchers to compare the observed probabilities for each value of the outcome to the model predictions for each outcome category. After generating the predicted probabilities using the `dpois` function, we can plot the predicted probabilities against the observed data:

```
pdat <- data.frame(Counts=rep(0:10,2),vals=c(dpois(0:10,mean(m
1$fitted.values)),table(X$num.children)/length(X$num.
children)),
                    type=rep(c("model","observed"),each=11))
ggplot(pdat,aes(x=Counts,y=vals,group=type,linetype=type)) +
  theme_bw() + geom_line() + geom_point() + labs(y="Predicted
Probability",linetype="") +
  theme(legend.position = "bottom")
```

As we can see in Figure 9.5, the predictions for three children on are fairly accurate, but our model underpredicts the probability of having no children, overpredicts the probability of having one child, and underpredicts the probability of having two children. We move on to other models for predicting counts to see if we can improve upon this original model in terms of fitting the observed data.

Negative Binomial Regression

The key assumption in the PRM is equidispersion, which rarely holds empirically. In practice, the variance of a count variable is often greater than its mean. There are a couple of common ways to address this assumption. First, we can just add extra dispersion to the Poisson regression. This is known as the quasi-Poisson regression. It can be fitted using `glm` with the family set to `quasipoisson`. However, R does not work well with quasi-likelihood models like these and fitting this model requires workarounds for simple tasks like calculating fit statistics.[1] A more common approach that is similar in concept is to use a negative binomial regression (NBRM).

The NBRM adds an error term to the PRM that is assumed to be uncorrelated with the independent variables in the model. As with our previous forays into the error term with categorical models, the error term is unobserved and thus we do not know its variance. But, if we assume that $\exp(\varepsilon)$ follows a gamma distribution and mix the gamma distribution with the Poisson distribution (Cameron and Trivedi 2013; Long 1997), we arrive at a negative binomial distribution.

Estimation

To fit a negative binomial, we use the glm.nb function from the MASS package:

```
m2 <- glm.nb(num.children ~ religious + minority  + female +
age + education + divo  rced + married + widow ,data=X)
summary(m2)

Call:
glm.nb(formula = num.children ~ religious + minority + female
+
    age + education + divorced + married + widow, data = X,
init.theta = 50.90766428,
    link = log)

Deviance Residuals:
    Min        1Q   Median        3Q       Max
-2.8150   -1.3083  -0.0538    0.5676    4.4833

Coefficients:
             Estimate Std. Error z value Pr(>|z|)
(Intercept) -0.421293   0.104660  -4.025 5.69e-05 ***
religious    0.010256   0.005935   1.728 0.083970 .
minority     0.178873   0.066138   2.705 0.006840 **
female       0.203370   0.035773   5.685 1.31e-08 ***
age          0.016453   0.001145  14.370  < 2e-16 ***
education   -0.016037   0.004922  -3.258 0.001121 **
divorced     0.171697   0.050252   3.417 0.000634 ***
married      0.287009   0.061646   4.656 3.23e-06 ***
widow       -0.065630   0.059871  -1.096 0.272995
---
Signif. codes:  0 '***' 0.001 '**' 0.01 '*' 0.05 '.' 0.1 ' ' 1

(Dispersion parameter for Negative Binomial(50.9077) family
taken to be 1)

    Null deviance: 3048.9  on 2156  degrees of freedom
Residual deviance: 2631.3  on 2148  degrees of freedom
  (47 observations deleted due to missingness)
AIC: 6781.1

Number of Fisher Scoring iterations: 1


            Theta:  50.9
        Std. Err.:  49.4

 2 x log-likelihood:  -6761.111
```

Observant readers may notice that the coefficients from our NBRM are almost exactly the same as those from our PRM. Indeed, the mean structure of the NBRM and PRM is the same. The addition of the error term does not change the predicted counts from our model at different levels of the independent variables but does change the standard error of those estimates as this term affects the modeled variance. In addition to the regular regression output, we see some additional output, namely, a theta estimate, its standard error, and 2×log-likelihood. These are reports on the additional error term we just added to the model. Users of other statistical packages may be more familiar with an alpha estimate rather than theta. Luckily for us, $\theta = 1/\alpha$ so a θ of 50.9 is equivalent to an α of 0.02 reported in other packages. Somewhat confusingly, *smaller* values of θ indicate *more* dispersion. Here, it might be more useful to think about the dispersion parameter in terms of α or $1/\theta$, which is part of the reason other packages report α instead of θ. When $\alpha=0$, the NBRM and PRM are equivalent. Thus, we can think of the PRM as a reduced model of the NBRM. Therefore, we can perform a likelihood ratio test to determine if our data are overdispersed:

```
> lr.test(m1,m2)
    LL.Full LL.Reduced G2.LR.Statistic DF p.value
1 -3381.115  -3380.556        1.117988  1 0.29035
```

As shown by our LR test and as one might expect given that $\alpha=0.02$ tested against a null of $\alpha=0$, there is not enough evidence to reject the null of equidispersion. Nevertheless, we walk through interpretations for the NBRM for cases where there is overdispersion. More confident readers may skip ahead because the interpretation for the NBRM is exactly the same as the interpretation for the PRM.

Incidence Rate Ratios

We again use list.coef to report both raw coefficients and IRRs. Using ggplot, we can plot the IRRs to show that all else equal, being married, female, a racial or ethnic minority, and divorced is associated with a 1.33 times, 1.23 times, 1.20 times, and 1.19 increase in the predicted number of children living at home, respectively. Each of these increases is statistically significant ($p<.001$). Similarly, but less drastically, each additional unit of age, education, and religiosity is associated with a 1.02, 0.98, and 1.01 times change in the predicted number of children (Figure 9.6).

```
list.coef(m2)
$out
```

	variables	b	SE	z	ll	ul	p.val	exp.b	ll.exp.b	ul.exp.b	percent	CI
1	(Intercept)	-0.421	0.105	-4.025	-0.626	-0.216	0.000	0.656	0.534	0.806	-34.380	95%
2	religious	0.010	0.006	1.728	-0.001	0.022	0.090	1.010	0.999	1.022	1.031	95%
3	minority	0.179	0.066	2.705	0.049	0.309	0.010	1.196	1.050	1.361	19.587	95%
4	female	0.203	0.036	5.685	0.133	0.273	0.000	1.226	1.143	1.315	22.553	95%
5	age	0.016	0.001	14.370	0.014	0.019	0.000	1.017	1.014	1.019	1.659	95%
6	education	-0.016	0.005	-3.258	-0.026	-0.006	0.002	0.984	0.975	0.994	-1.591	95%
7	divorced	0.172	0.050	3.417	0.073	0.270	0.001	1.187	1.076	1.310	18.732	95%
8	married	0.287	0.062	4.656	0.166	0.408	0.000	1.332	1.181	1.504	33.244	95%
9	widow	-0.066	0.060	-1.096	-0.183	0.052	0.219	0.936	0.833	1.053	-6.352	95%

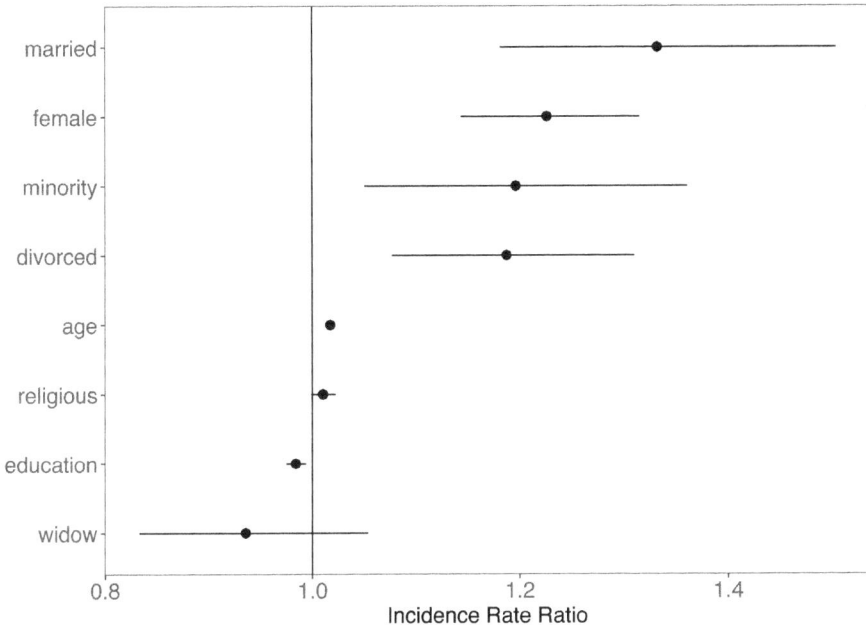

FIGURE 9.6
Sorted plot of incidence rate ratios.

Marginal Effects

We can also examine the marginal effects of variables using the `margins` functions. Take, for example, marital status. We can call `margins.des` to create design matrices for each of the four possible combinations of marital status using our three indicator variables, combine them using `rbind`, and then use the resulting matrix as input for `margins.dat` to create predictions, which we can plot:

```
d1 <- margins.des(m2,ivs=expand.grid(divorced=0,married=0,widow=0))
d2 <- margins.des(m2,ivs=expand.grid(divorced=1,married=0,widow=0))
d3 <- margins.des(m2,ivs=expand.grid(divorced=0,married=1,widow=0))
d4 <- margins.des(m2,ivs=expand.grid(divorced=0,married=0,widow=1))
design <- rbind(d1,d2,d3,d4)
design
pdat<- margins.dat(m2,design)
pdat <- mutate(pdat,marital=c("Single","Divorced","Married","Widow"))
ggplot(pdat,aes(x=marital,y=fitted,ymin=ll,ymax=ul)) +
  theme_bw() + geom_pointrange() + labs(x="",y="Predicted # of
Children")
```

As shown in Figure 9.7, being married or divorced is associated with having significantly more children living at home than being single or widowed. We

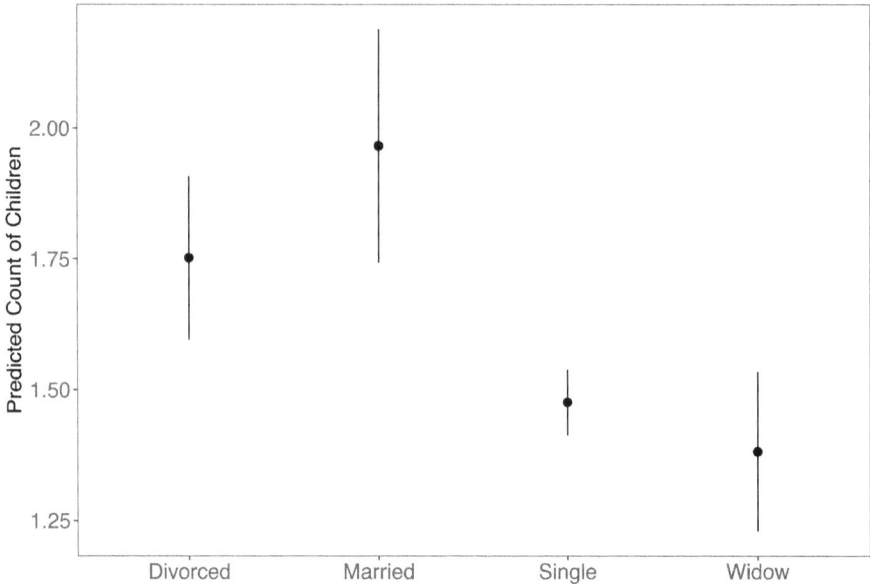

FIGURE 9.7
Predicted number of children by marital status.

can use `first.diff.fitted` to confirm the statistical significance of these differences:

```
pdat<- margins.dat(m1,design)
first.diff.fitted(m1,pdat,compare=c(1,2))
first.diff.fitted(m1,pdat,compare=c(1,3))
first.diff.fitted(m1,pdat,compare=c(1,4))

  first.diff std.error statistic p.value     ll     ul
1    -0.276     0.086    -3.227   0.001 -0.444 -0.109
> first.diff.fitted(m2,pdat,compare=c(1,3))
  first.diff std.error statistic p.value     ll     ul
1    -0.491     0.118     -4.16       0 -0.722 -0.259
> first.diff.fitted(m2,pdat,compare=c(1,4))
  first.diff std.error statistic p.value     ll     ul
1     0.094     0.083     1.123   0.262  -0.07  0.257
```

We can also use `marginaleffects` to show the AME of each variable:

```
summary(marginaleffects(m2))

       Term  Effect Std. Error z value   Pr(>|z|)     2.5 %   97.5 %
1 religious 0.01667   0.009652   1.727 0.08421752 -0.002251  0.03558
2  minority 0.29068   0.107614   2.701 0.00691007  0.079762  0.50160
```

```
3     female  0.33049   0.058443    5.655 1.5586e-08   0.215946  0.44504
4        age  0.02674   0.001937   13.807 < 2.22e-16   0.022943  0.03053
5 education -0.02606   0.008021   -3.249 0.00115712  -0.041781 -0.01034
6   divorced  0.27902   0.081804    3.411 0.00064768   0.118686  0.43935
7    married  0.46641   0.100518    4.640 3.4825e-06   0.269401  0.66342
8      widow -0.10665   0.097320   -1.096 0.27312964  -0.297396  0.08409
```

Predicted Probabilities

Finally, we generate and plot predicted probabilities for each count to see how our model is performing relative to the observed data. Here, unlike with the PRM, we use the dnbinom function to generate predicted probabilities. Unlike the Poisson distribution, dnbinom requires two parameters: a θ parameter (saved in our model object) as well as a μ (mean). For example, the following call gives us the predicted probability of having no children (Figure 9.8):

```
dnbinom(0,m2$theta,mu=mean(m2$fitted.values))
[1] 0.2019653
```

We can also generate predicted probabilities for each count up to 10 and plot them using the following call:

```
pdat <- data.frame(Counts=rep(0:10,2),vals=c(dnbinom(0:10,
m2$theta,mu=mean(m2$fitted.values)),table(X$num.children)/
length(X$num.children)),
                type=rep(c("model","observed"),each=11))
ggplot(pdat,aes(x=Counts,y=vals,group=type,linetype=type)) +
  theme_bw() + geom_line() + geom_point()
```

Unfortunately, accounting for overdispersion does not improve the fit of our model. This is perhaps due to there being multiple sources of zeros leading to a relatively high proportion of zero children relative to what we would expect given the proportion of people who report only having one child. Indeed, it seems like our curve could be fit to the 1–10 counts easily if there is a different way to model the 0 count. The next set of models attempts to do this.

Zero-Inflated Models

Zero-inflated models further expand on the PRM and NBRM by allowing different data-generating processes to create zeros (Lambert 1992; Mullahy 1986). Depending on whether we were trying to improve upon the PRM and NBRM, we can fit a corresponding ZIP regression or ZINB regression. Both assume that there are two distinct sources of zeros for a given sample. The first, called *structural zeros*, are people who will always be zero. In our case,

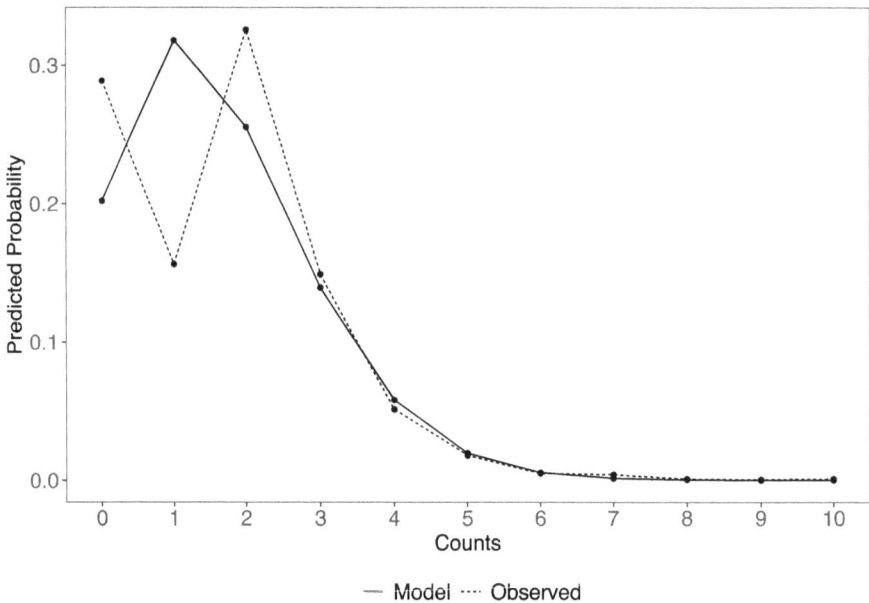

FIGURE 9.8
Predicted probabilities compared to observed proportions.

they can be people who cannot and will not have children. The second, called sampling zeros, are people who could one day have nonzero counts, they just happen to have been sampled into the study when they had a zero count. These may be people who want children but have not yet done so. The model is fit in three steps.

First, we use a BRM to predict whether someone is a structural zero or a sampling zero. Second, we use a ZIP and ZINB to predict the count for everyone who is not a structural zero. Third, we mix the two probabilities together. Expected counts then are:

$$E(y_i | \mathbf{x}_i, \mathbf{z}_i) = \psi_i \times 0 + (1 - \psi_i) \times \mu_i \qquad (9.6)$$

where x_i is the vector of predictors for our ZIP or ZINB, z_i is the vector of predictors for our BRM, ψ_i is the predicted probability of being a structural zero (derived from a binary logit or probit), and μ_i is the predicted count from our ZIP or ZINB. In other words, the ZIP or ZINB is a weighted average of a BRM and a count model. Because it is a mixture of two separate models, it is possible to have a different set of predictors for the binary portion and for the count portion. Indeed, there are often different predictors for structural zeros than for the value of a count response. For simplicity, however, we specify a model that has the same set of predictors.

Estimation

Zero-inflated models can be fit using the `zeroinfl` function in the `pscl` package (Jackman et al. 2015). Below, we fit two models, m3 and m4, which only differ in whether we specify the option, dist="poisson" or dist="negbin," for a ZIP and ZINB, respectively.

```
m3 <- zeroinfl(num.children ~ religious + minority  + female +
age + education + divorced + married + widow | religious +
minority  + female + age + education + divorced + married +
widow, dist = "poisson", data = X)

m4 <- zeroinfl(num.children ~ religious + minority  + female +
age + education + divorced + married + widow | religious +
minority  + female + age + education + divorced + married +
widow, dist = "negbin", data = X)
```

Walking through the structure of the function, we see two sets of variables separated by a | divider. Otherwise, the call is similar to our other function calls. For zero-inflated models, the first set of variables are our *x* variables, those that predict the count for sampling zeros. The second set of variables are our *z* variables, those that predict whether someone is a structural zero or a sampling zero. Here, we specify the exact same set of variables, but it is possible and perhaps with stronger theory there would be a good reason to specify different variables predicting whether someone has a child at all compared to how many children they have. We then call summary(m4) to produce the output for the model. Note that we will use m4 because it is the full model. Model 3 is a reduced version of m4. All interpretations are the same except without the additional dispersion parameter if we wanted to interpret our ZIP instead:

```
summary(m4)

Call:
zeroinfl(formula = num.children ~ religious + minority +
female + age + education + divorced + married +
    widow | religious + minority + female + age + education +
divorced + married + widow, data = X, dist = "negbin")

Pearson residuals:
    Min        1Q     Median        3Q        Max
-1.76945  -0.63846  -0.04658   0.50398   6.25580

Count model coefficients (negbin with log link):
            Estimate Std. Error z value Pr(>|z|)
(Intercept) 0.271772   0.116872   2.325 0.020052 *
religious   0.011615   0.005999   1.936 0.052872 .
minority    0.231163   0.067511   3.424 0.000617 ***
female      0.115754   0.037112   3.119 0.001814 **
age         0.007099   0.001345   5.278 1.31e-07 ***
```

```
education   -0.018016   0.004961  -3.631 0.000282 ***
divorced     0.135940   0.049515   2.745 0.006043 **
married      0.218896   0.060750   3.603 0.000314 ***
widow        0.049593   0.059602   0.832 0.405368
Log(theta)  16.638498   7.324077   2.272 0.023101 *

Zero-inflation model coefficients (binomial with logit link):
            Estimate Std. Error z value Pr(>|z|)
(Intercept)  5.68720    0.97040   5.861 4.61e-09 ***
religious   -0.06258    0.05650  -1.108  0.26800
minority     1.11345    0.48157   2.312  0.02077 *
female      -1.51404    0.32240  -4.696 2.65e-06 ***
age         -0.26785    0.03800  -7.048 1.81e-12 ***
education    0.17876    0.05537   3.229  0.00124 **
divorced    -0.32305    1.41747  -0.228  0.81972
married     -1.83854    1.55586  -1.182  0.23733
widow        4.22702    1.37816   3.067  0.00216 **
---
Signif. codes:  0 '***' 0.001 '**' 0.01 '*' 0.05 '.' 0.1 ' ' 1

Theta = 16827044.0231
Number of iterations in BFGS optimization: 31
Log-likelihood: -3258 on 19 Df
```

In the above output, we see a humorously large θ, indicating that our $\alpha = 1/\theta$ is approaching zero. This suggests that there is equidispersion and that we do not need to waste the extra degree of freedom estimating the dispersion parameter and can instead stick to a ZIP. Nevertheless, the results between the ZINB and ZIP in this case are going to be virtually identical and we move onto interpreting the more complicated model. Indeed, we can confirm this with an LR test showing that the additional parameter does not significantly improve the fit of the model:

```
> lr.test(m3,m4)
     LL.Full  LL.Reduced  G2.LR.Statistic DF p.value
1 -3257.547   -3257.547      1.062857e-05  1  0.9974
```

Interpretation

We can call list.coef to interpret the IRRs for our model. Like our regression output, we have two sets of results to interpret, one from the count model and one from the BRM. Here, the count model would be interpreted similarly to how we have done in the previous sections. For the BRMs, the results in the exp.b column are odds ratios.

```
list.coef(m4)
$out
```

	variables	b	SE	z	ll	ul	p.val	exp.b	ll.exp.b	ul.exp.b	percent	CI
1	count_(Intercept)	0.272	0.117	2.325	0.043	0.501	0.027	1.312	1.044	1.650	31.230	95 %
2	count_religious	0.012	0.006	1.936	0.000	0.023	0.061	1.012	1.000	1.024	1.168	95 %
3	count_minority	0.231	0.068	3.424	0.099	0.363	0.001	1.260	1.104	1.438	26.006	95 %
4	count_female	0.116	0.037	3.119	0.043	0.188	0.003	1.123	1.044	1.207	12.272	95 %
5	count_age	0.007	0.001	5.278	0.004	0.010	0.000	1.007	1.004	1.010	0.712	95 %
6	count_education	-0.018	0.005	-3.631	-0.028	-0.008	0.001	0.982	0.973	0.992	-1.785	95 %
7	count_divorced	0.136	0.050	2.746	0.039	0.233	0.009	1.146	1.040	1.262	14.562	95 %
8	count_married	0.219	0.061	3.603	0.100	0.338	0.001	1.245	1.105	1.402	24.470	95 %
9	count_widow	0.050	0.060	0.832	-0.067	0.166	0.282	1.051	0.935	1.181	5.084	95 %
10	zero_(Intercept)	5.687	0.970	5.861	3.785	7.589	0.000	294.995	44.039	1976.010	29399.482	95 %
11	zero_religious	-0.063	0.056	-1.108	-0.173	0.048	0.216	0.939	0.841	1.049	-6.065	95 %
12	zero_minority	1.113	0.482	2.312	0.170	2.057	0.028	3.045	1.185	7.824	204.455	95 %
13	zero_female	-1.514	0.322	-4.696	-2.146	-0.882	0.000	0.220	0.117	0.414	-77.996	95 %
14	zero_age	-0.268	0.038	-7.048	-0.342	-0.193	0.000	0.765	0.710	0.824	-23.497	95 %
15	zero_education	0.179	0.055	3.229	0.070	0.287	0.002	1.196	1.073	1.333	19.573	95 %
16	zero_divorced	-0.323	1.418	-0.228	-3.102	2.455	0.389	0.724	0.045	11.649	-27.632	95 %
17	zero_married	-1.838	1.556	-1.182	-4.887	1.211	0.198	0.159	0.008	3.356	-84.087	95 %
18	zero_widow	4.226	1.378	3.066	1.525	6.928	0.004	68.460	4.594	1020.195	6746.028	95 %

To ease interpretation, we sort and plot the IRRs and ORs in Figure 9.9:

Note that we removed the odds ratio for widowed from the plot because the range for the confidence interval dwarfs the scale of the other variables, so it is difficult to visualize the results. As shown in Figure 9.9, once we take into account the difference between structural and sampling zeros, being a racial or ethnic minority has a larger positive effect on the number of children living at home than in our baseline NBRM. Being a racial minority is now associated with a 26 percent increase in the predicted number of children. Note, however, that being a racial minority is also associated with a 204 percent increase in the odds of being a structural zero, that is, never having children. Substantively, we would conclude something along the lines of being a racial minority is associated with not having children, but among those with children, a significantly higher number of children live at home. Otherwise, the effects of the other variables are similar to those in our baseline NBRM model. Compared to being single, being married or divorced is associated with having more children at home. Female respondents report having more children at home than male respondents. Higher levels of religiosity are associated with more children, as is being older. Having more years of education is associated with having fewer children. As with our previous examples, we can plot the predicted probability across the range of a variable like education (Figure 9.10):

```
design <- margins.des(m4,ivs=expand.grid(education=10:18))
pdat <- margins.dat(m4,design,pscl.data=X)
ggplot(pdat,aes(x=education,y=fitted,ymin=ll,ymax=ul)) +
  theme_bw() + geom_point() + geom_line() + geom_
ribbon(alpha=.1) +
  labs(x="Education",y="Predicted Count of Children")
```

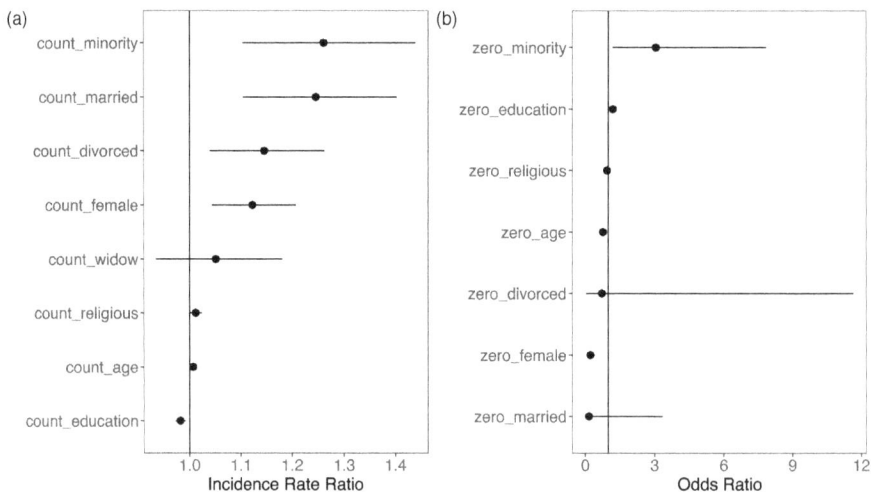

FIGURE 9.9
IRR (a) and OR (b) from a ZINB model.

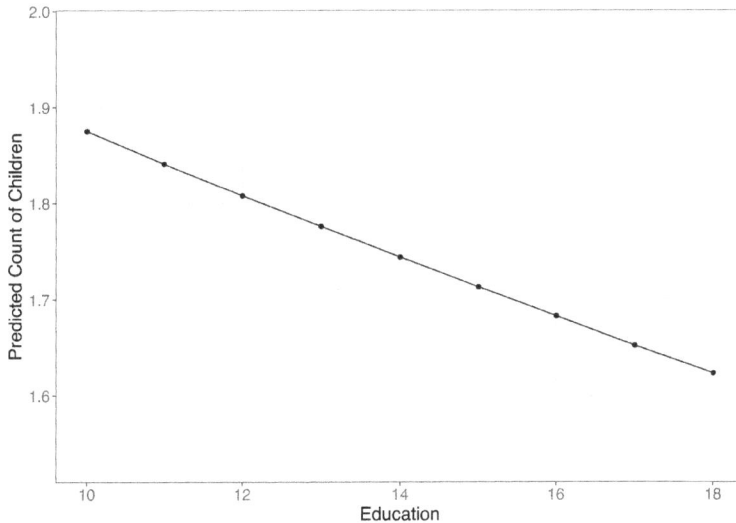

FIGURE 9.10
Predicted number of children by years of education.

Although the call for creating the design matrix is the same, an important difference in our `margins.dat` call for zero-inflated models is the use of the `pscl.dat=X` option, which enables `margins.dat` to compute the means of covariates. Put differently, we need to give the original data used in the model statement because it is not included with the model object itself.

It is also important to note that `first.diff.fitted` and `second.diff.fitted` only support nonparametric inference (i.e., bootstrapping) for zero-inflated models because the Delta method function does not work with `pscl` model objects. A workaround using simulations rather than the Delta method is to use the `compare.margins` function. The following call tests whether the difference in the predicted counts at 10 versus 18 years of schooling are different, with covariates set to their means, that is, MEMs:

```
> design <- margins.des(m4,expand.grid(education=c(10,18)))
> pdat <- margins.dat(m4,design,pscl.data=X)
> compare.margins(margins=pdat$fitted,margins.ses=pdat$se)
$difference
[1] 0.252

$p.value
[1] 0
```

As shown in the output, those with 10 years of schooling are predicted to have .25 more children living at home than those with 18 years of schooling, and this difference is significant. Similarly, suppose we wanted to know the conditional MEM of being female at the lowest compared to highest levels of education. The following code generates those results:

```
> design <- margins.des(m4,expand.grid(female=c(1,0),education=c(10,18)))

> pdat <- margins.dat(m4,design,pscl.data=X)

> pdat
  female education religious minority   age divorced married widow fitted    se    ll    ul
1      1        10     3.603    0.076 53.238    0.121   0.067 0.096  1.975 0.062 1.855 2.096
2      0        10     3.603    0.076 53.238    0.121   0.067 0.096  1.759 0.065 1.632 1.887
3      1        18     3.603    0.076 53.238    0.121   0.067 0.096  1.710 0.054 1.605 1.815
4      0        18     3.603    0.076 53.238    0.121   0.067 0.096  1.523 0.056 1.413 1.634

> compare.margins(margins=pdat$fitted[1:2],margins.ses=pdat$se[1:2])
$difference
[1] 0.216

$p.value
[1] 0.008

> compare.margins(margins=pdat$fitted[3:4],margins.ses=pdat$se[3:4])
$difference
[1] 0.187

$p.value
[1] 0.008
```

Above we see that the difference between men and women is slightly smaller with more education (.216 vs. .187). We can produce similar estimates using AMEs instead of MEMs. Specifically, we use marginaleffects to generate the conditional AMEs and then use the compare.margins function to test for differences between the estimates.

```
> ma1<-summary(marginaleffects(m4,variables="female",newdata=d
atagrid(education=10)))
> ma2<-summary(marginaleffects(m4,variables="female",newdata=d
atagrid(education=18)))
> ma<-rbind(ma1,ma2)
> ma
    Term Effect Std. Error z value  Pr(>|z|)   2.5 % 97.5 %
1 female 0.2184    0.06880   3.175 0.0015003 0.08357 0.3533
2 female 0.1930    0.05917   3.262 0.0011050 0.07706 0.3090

> compare.margins(margins=ma$estimate,margins.ses=ma$std.
error)
$difference
[1] 0.025

$p.value
[1] 0.387
```

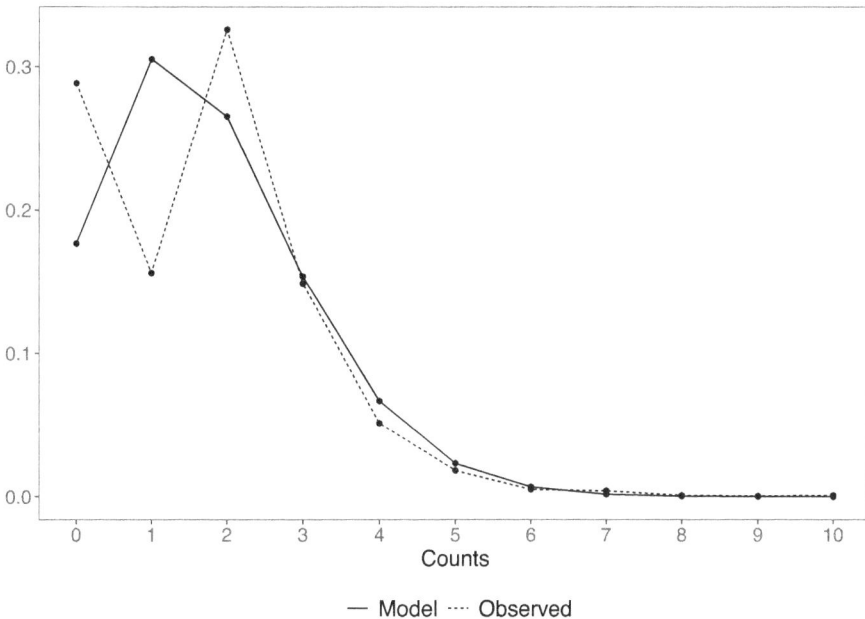

FIGURE 9.11
Predicted probabilities compared to observed proportions.

As shown in the output, the AME of being female is 0.218 at 10 years of education and 0.193 at 18 years, with the difference of 0.025 being not statistically significant with a *p*-value of 0.387.

Finally, we can calculate and plot the predicted probability of each level of the outcome. As shown in Figure 9.11, we can compare the predicted probabilities from our ZINB to the observed proportion of respondents at each count. Here, unfortunately, we do not improve on the fit of the data, even when accounting for multiple sources of zeros.

```
> pdat <- data.frame(Counts=rep(0:10,2),vals=c(predict(m4,newd
ata=design,type="prob"),table(X$num.children)/length(X$num.
children)),
+                         type=rep(c("model","observed"),each=11))
> ggplot(pdat,aes(x=Counts,y=vals,group=type,linetype=type)) +
+    theme_bw() + geom_line() + geom_point() + scale_x_
continuous(breaks=0:10) +
+    theme(legend.position="bottom") + labs(linetype="")
```

Comparing Count Models

So far, we have examined model fit by plotting predicted probabilities to observed proportions. These "eyeball" tests are instructive but informal. With overdispersion, an LR test as we have done above provides a formal test of whether the additional dispersion parameter is warranted. Unfortunately, we cannot perform an LR test comparing zero-inflated models to baseline PRM or NBRM because these models are not nested within one another. To more formally test these models, we can use a Vuong (1989) test for non-nested models. In the Vuong test, we are comparing the predicted probabilities from two nonnested models:

$$\widehat{Pr}_1(y_i \mid \boldsymbol{x}_i) \text{ and } \widehat{Pr}_2(y_i \mid \boldsymbol{x}_i) \tag{9.7}$$

Vuong (1989) defines m_i as $\ln\left[\widehat{Pr}_1\left(y_i \mid x_i\right)\right] - \ln[\widehat{Pr}_2(y_i \mid x_i)]$. Given this, the Vuong test statistic is:

$$V = \frac{\sqrt{N}\bar{m}}{s_m} \tag{9.8}$$

which asymptotically follows a normal distribution. Under the null hypothesis, V=0. A large and significant positive V statistic indicates that the first model is preferred, whereas a large and significant negative V statistic indicates that the second model is preferred. The count.fit function automates the testing and reporting of the Vuong test as well as other comparisons:

```
cf <- count.fit(m1,y.range=0:10)
Vuong Non-Nested Hypothesis Test-Statistic:
(test-statistic is asymptotically distributed N(0,1) under the
 null that the models are indistinguishable)
------------------------------------------------------------
            Vuong z-statistic          H_A    p-value
Raw                 -8.052610 model2 > model1 4.0523e-16
AIC-corrected       -7.466102 model2 > model1 4.1303e-14
BIC-corrected       -5.801453 model2 > model1 3.2871e-09
Vuong Non-Nested Hypothesis Test-Statistic:
(test-statistic is asymptotically distributed N(0,1) under the
 null that the models are indistinguishable)
------------------------------------------------------------
            Vuong z-statistic          H_A    p-value
Raw                 -8.197698 model2 > model1 < 2.22e-16
AIC-corrected       -7.597909 model2 > model1 1.5048e-14
BIC-corrected       -5.895566 model2 > model1 1.8670e-09
```

To use count.fit, we specify a baseline PRM (m1) and a range of the outcome to calculate predicted probabilities (0–10 children). count.fit produces two sets of output, the first comparing the PRM to the ZIP and the second comparing the NBRM to the ZINB. The results show a "raw" Vuong statistic as well as two finite sample corrections based on the AIC and BIC. In all cases, the zero-inflated models are preferred based on the significant and negative test statistics. In addition, count.fit allows for comparisons based on AIC and BIC and plots of the coefficients as well as the residuals.

```
cf$ic
       Poisson Neg Binom      ZIP       ZNB
BIC 6831.317  6837.876 6653.270 6660.947
AIC 6780.229  6781.111 6551.094 6553.094
```

Calling cf$ic returns the BIC and AIC for all four models. Based on both the BIC and AIC, the ZIP is the preferred model with the lowest BIC and the lowest AIC. Calling cf$models, as we have done below, returns all regression coefficients together, making comparisons simpler. As illustrated below, this returns the coefficient, the standard error, the z-statistic, and the p-value for each coefficient. Each model-type is abbreviated as follows: "P" denotes Poisson regression, "NB" denotes negative binomial, "ZIP" denotes zero-inflated Poisson, and "ZNB" denotes zero-inflated negative binomial.

cf$models

	Pcoef	Pse	Pz	Ppval	NBcoef	NBse	NBz	NBpval	ZIPcoef	ZIPse	ZIPz	ZIPpval	ZNBcoef	ZNBse
count_(Intercept)	-0.415	0.103	-4.027	0.000	-0.421	0.105	-4.025	0.000	0.272	0.117	2.325	0.027	0.272	0.117
count_religious	0.010	0.006	1.740	0.088	0.010	0.006	1.728	0.090	0.012	0.006	1.936	0.061	0.012	0.006
count_minority	0.180	0.065	2.765	0.009	0.179	0.066	2.705	0.010	0.231	0.068	3.424	0.001	0.231	0.068
count_female	0.202	0.035	5.748	0.000	0.203	0.036	5.685	0.000	0.116	0.037	3.119	0.003	0.116	0.037
count_age	0.016	0.001	14.502	0.000	0.016	0.001	14.370	0.000	0.007	0.001	5.278	0.000	0.007	0.001
count_education	-0.016	0.005	-3.297	0.002	-0.016	0.005	-3.258	0.002	-0.018	0.005	-3.631	0.001	-0.018	0.005
count_divorced	0.172	0.049	3.495	0.001	0.172	0.050	3.417	0.001	0.136	0.050	2.746	0.009	0.136	0.050
count_married	0.286	0.060	4.733	0.000	0.287	0.062	4.656	0.000	0.219	0.061	3.603	0.001	0.219	0.061
count_widow	-0.062	0.059	-1.065	0.226	-0.066	0.060	-1.096	0.219	0.050	0.060	0.832	0.282	0.050	0.060
zero_(Intercept)	0.000	0.000	NaN	NaN	0.000	0.000	NaN	NaN	5.687	0.970	5.861	0.000	5.687	0.970
zero_religious	0.000	0.000	NaN	NaN	0.000	0.000	NaN	NaN	-0.063	0.056	-1.108	0.216	-0.063	0.056
zero_minority	0.000	0.000	NaN	NaN	0.000	0.000	NaN	NaN	1.113	0.482	2.312	0.028	1.113	0.482
zero_female	0.000	0.000	NaN	NaN	0.000	0.000	NaN	NaN	-1.514	0.322	-4.696	0.000	-1.514	0.322
zero_age	0.000	0.000	NaN	NaN	0.000	0.000	NaN	NaN	-0.268	0.038	-7.048	0.000	-0.268	0.038
zero_education	0.000	0.000	NaN	NaN	0.000	0.000	NaN	NaN	0.179	0.055	3.229	0.002	0.179	0.055
zero_divorced	0.000	0.000	NaN	NaN	0.000	0.000	NaN	NaN	-0.323	1.418	-0.228	0.389	-0.323	1.418
zero_married	0.000	0.000	NaN	NaN	0.000	0.000	NaN	NaN	-1.838	1.556	-1.182	0.198	-1.838	1.556
zero_widow	0.000	0.000	NaN	NaN	0.000	0.000	NaN	NaN	4.226	1.378	3.066	0.004	4.226	1.378

	ZNBz	ZNBpval
count_(Intercept)	2.325	0.027
count_religious	1.936	0.061
count_minority	3.424	0.001
count_female	3.119	0.003
count_age	5.278	0.000
count_education	-3.631	0.001
count_divorced	2.746	0.009
count_married	3.603	0.001
count_widow	0.832	0.282
zero_(Intercept)	5.861	0.000
zero_religious	-1.108	0.216
zero_minority	2.312	0.028
zero_female	-4.696	0.000
zero_age	-7.048	0.000
zero_education	3.229	0.002
zero_divorced	-0.228	0.389
zero_married	-1.182	0.198
zero_widow	3.066	0.004

Calling cf$models.pic produces a graphical comparison of the model coefficients, as illustrated in Figure 9.12.

Finally, calling cf$pic produces an "observed–predicted" plot showing how each model performs in terms of fitting the observed data. This is illustrated

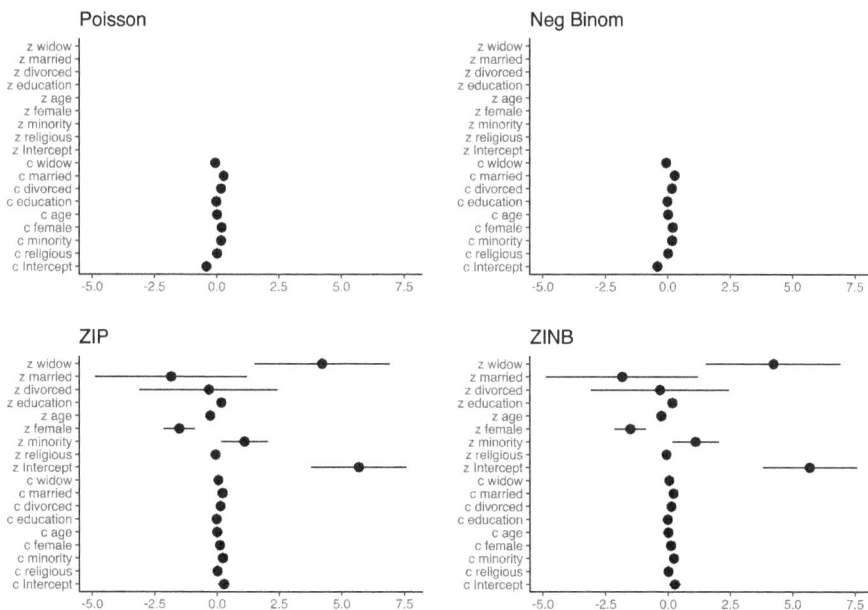

FIGURE 9.12
Comparison of coefficients. Plot generated by the count.fit function.

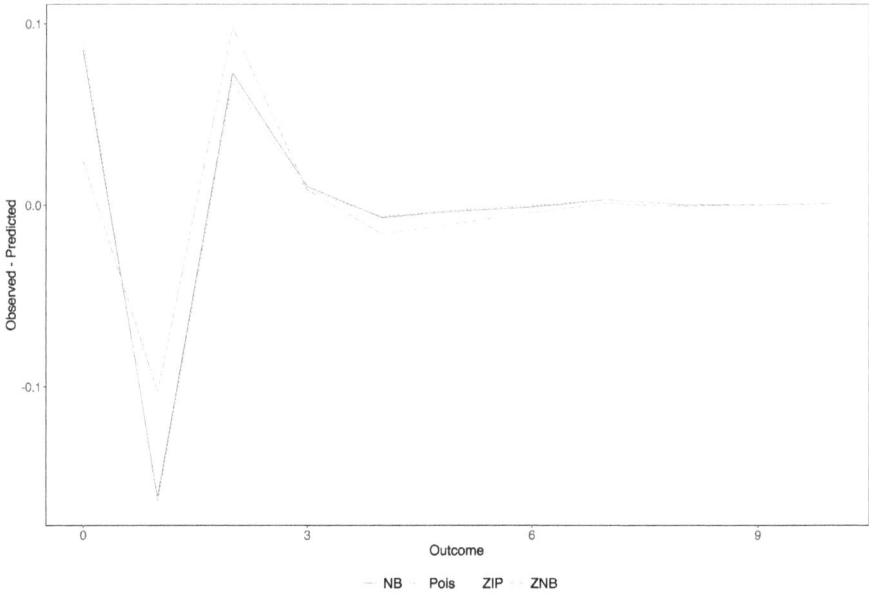

FIGURE 9.13
Observed–predicted plot. Plot generated by the `count.fit` function.

in Figure 9.13. Here, we can see that the Poisson and negative binomial versions of whatever model we choose fit the data almost exactly the same. This makes sense given the nonsignificant dispersion parameter. The zero-inflated versions of our models do a better job predicting zeros, unsurprisingly, and a better job predicting ones, but a worse job predicting higher counts.

Truncated Counts

Two more classes of models are often used to model count data. First, we address models for truncated count distributions and then we discuss models that treat zeros and counts as distinct processes (i.e., hurdle models). One common issue is that zeros often represent missing data in the real world. Suppose we collected data on the number of children from a parent's group. We know that there are childless people in the world, but we can only access our data from parents, so our data necessarily include only people with at least one child. Zero-truncated models are designed to deal with exactly these cases. We start with a PRM:

$$\Pr(y_i|\boldsymbol{x}) = \frac{e^{-\mu_i}\mu_i^{y_i}}{y_i!} \tag{9.9}$$

where $\mu_i = \exp(x_i\beta)$. Therefore, the conditional probability that $y_i = 0$ reduces to just:

$$\Pr(y_i = 0|x) = \exp(-\mu_i) \tag{9.10}$$

and the remaining probabilities is just $1 - \Pr(y_i = 0|x)$. Given that:

$$\Pr(A|B) = \frac{\Pr(A \cap B)}{\Pr(B)} \tag{9.11}$$

The conditional probability of a truncated count where we know that the outcome has no zeros ($\Pr(B)$) is:

$$\Pr y_i = ky_i > 0, x_i = \frac{\Pr(y_i = k|x_i)}{1 - exp(-\mu_i)} \tag{9.12}$$

Because $\Pr(y_i = k|x_i) = exp(x_i\beta) = \mu_i$, we can use the truncated model to generate a "conditional" and an "unconditional" set of predictions where the conditional probability is:

$$\Pr(y_i = ky_i > 0, x_i) = \frac{\mu_i}{1 - exp(-\mu_i)} \tag{9.13}$$

And the unconditional probability is:

$$\Pr(y_i = k|x_i) = exp(x_i\beta) \tag{9.14}$$

Conditional here refers to being conditional on the count being nonzero and positive. The "unconditional" predicted probability is still conditioning on the covariates. In addition, we can have a truncated negative binomial model as well, with the same underlying math and an additional dispersion parameter.

Estimation and Interpretation

To fit our truncated model, let us artificially recode zero children in our data as missing:

```
X <- mutate(X, ztkids=num.children)
X$ztkids[which(X$ztkids==0)]<-NA
```

Then, using the zerotrunc function from the countreg package, we can fit a ZTP and ZTNB model, respectively. As with our zero-inflated models, the only difference in the call is the dist option being "poisson" or "negbin":

```
m6 <- zerotrunc(ztkids ~ religious + minority  + female + age
+ education + divorced + married + widow, dist = "poisson",
data = X)

m7 <- zerotrunc(ztkids ~ religious + minority  + female + age
+ education + divorced + married + widow, dist = "negbin",
data = X)
summary(m7)

Call:
zerotrunc(formula = ztkids ~ religious + minority + female +
age + education + divorced + married + widow,
    data = X, dist = "negbin")

Deviance residuals:
    Min      1Q  Median      3Q     Max
-1.9332 -0.5270 -0.1574  0.3780  4.1108

Coefficients (truncated negbin with log link):
             Estimate Std. Error z value Pr(>|z|)
(Intercept)  0.272149   0.132974   2.047  0.04069 *
religious    0.006699   0.007083   0.946  0.34427
minority     0.219444   0.076202   2.880  0.00398 **
female       0.054257   0.043186   1.256  0.20899
age          0.008924   0.001486   6.007 1.89e-09 ***
education   -0.016045   0.005870  -2.733  0.00627 **
divorced     0.040158   0.059047   0.680  0.49644
married      0.199472   0.069947   2.852  0.00435 **
widow       -0.034926   0.069513  -0.502  0.61535
Log(theta)  15.430262  13.148036   1.174  0.24056
---
Signif. codes:  0 '***' 0.001 '**' 0.01 '*' 0.05 '.' 0.1 ' ' 1

Theta = 5026637.5068
Number of iterations in BFGS optimization: 43
Log-likelihood: -2171 on 10 Df
```

The output for a ZTNB is very similar to our baseline NBRM. As with our
NBRM, the output reports a θ estimate. We can use an LR test to confirm that
this large θ suggests that we do not have overdispersion. We can use list.
coef and the margins functions to interpret the ZTNB the same way we
interpret the NBRM. In this case, the functions are reporting IRRs for the
unconditional estimates, correcting for the zero truncation (Figure 9.14).

```
list.coef(m7)
$out
   variables      b    SE      z      ll      ul  p.val  exp.b ll.exp.b ul.exp.b  percent    CI
1 (Intercept) 0.272 0.133  2.047  0.012  0.533  0.049  1.313    1.012    1.704   31.278  95 %
2   religious 0.007 0.007  0.946 -0.007  0.021  0.255  1.007    0.993    1.021    0.672  95 %
3    minority 0.219 0.076  2.880  0.070  0.369  0.006  1.245    1.073    1.446   24.538  95 %
4      female 0.054 0.043  1.256 -0.030  0.139  0.181  1.056    0.970    1.149    5.576  95 %
5         age 0.009 0.001  6.007  0.006  0.012  0.000  1.009    1.006    1.012    0.896  95 %
6   education -0.016 0.006 -2.733 -0.028 -0.005  0.010  0.984    0.973    0.995   -1.592  95 %
7    divorced 0.040 0.059  0.680 -0.076  0.156  0.317  1.041    0.927    1.169    4.098  95 %
8     married 0.199 0.070  2.852  0.062  0.337  0.007  1.221    1.064    1.400   22.076  95 %
9       widow -0.035 0.070 -0.502 -0.171  0.101  0.352  0.966    0.843    1.107   -3.432  95 %
```

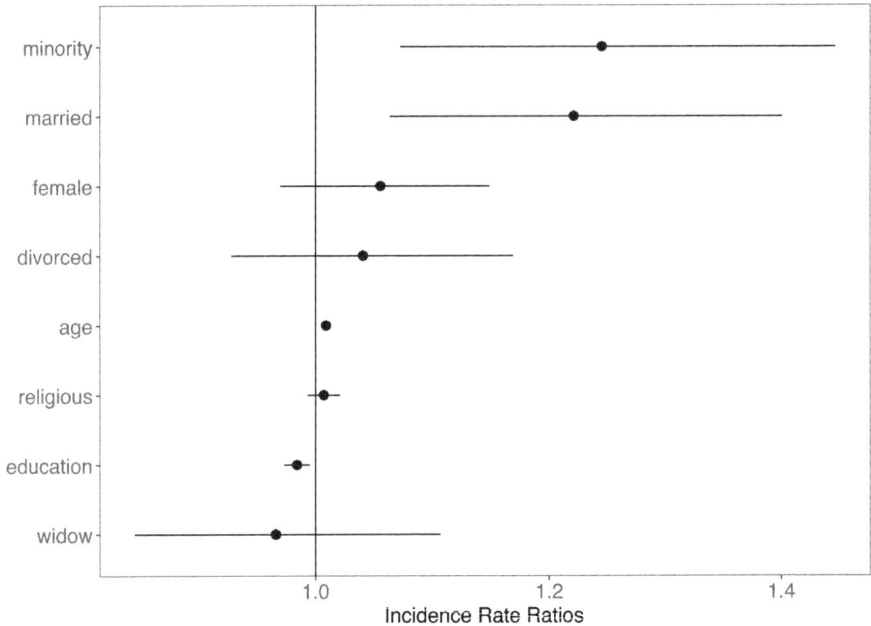

FIGURE 9.14
Incidence rate ratios from zero-truncated negative binomial regression.

We can also generate predicted counts. Unfortunately, only nonparametric predictions based on bootstrapping are possible at this time with truncated models. When working with zero-truncated models, `margins.dat` will rely on bootstrapped inference by default. The default settings entail taking 90% of the sample, with replacement, to generate model estimates, and this is repeated 1,000 times. Both the % of the sample and the number of repetitions is variable. We use these default settings to generate Figure 9.15:

```
d1 <- margins.des(m7,ivs=expand.grid(divorced=0,married=0,widow=0))
d2 <- margins.des(m7,ivs=expand.grid(divorced=1,married=0,widow=0))
d3 <- margins.des(m7,ivs=expand.grid(divorced=0,married=1,widow=0))
d4 <- margins.des(m7,ivs=expand.grid(divorced=0,married=0,widow=1))
design <- rbind(d1,d2,d3,d4)
design
pdat<- margins.dat(m7,design,num.sample=1000)
pdat <- mutate(pdat,type=c("Single","Divorced","Married","Widow"))
ggplot(pdat,aes(x=type,y=fitted,ymin=ll,ymax=ul)) +
  theme_bw() + geom_pointrange() + labs(x="",y="Predicted Count of
Children") +
  scale_y_continuous(limits=c(.5,3))
```

As shown in the figure, after correcting for the zero truncation, we expect single respondents to have 2.22 children compared to divorced respondents (2.28), married respondents (2.55), and widowed respondents (2.17).

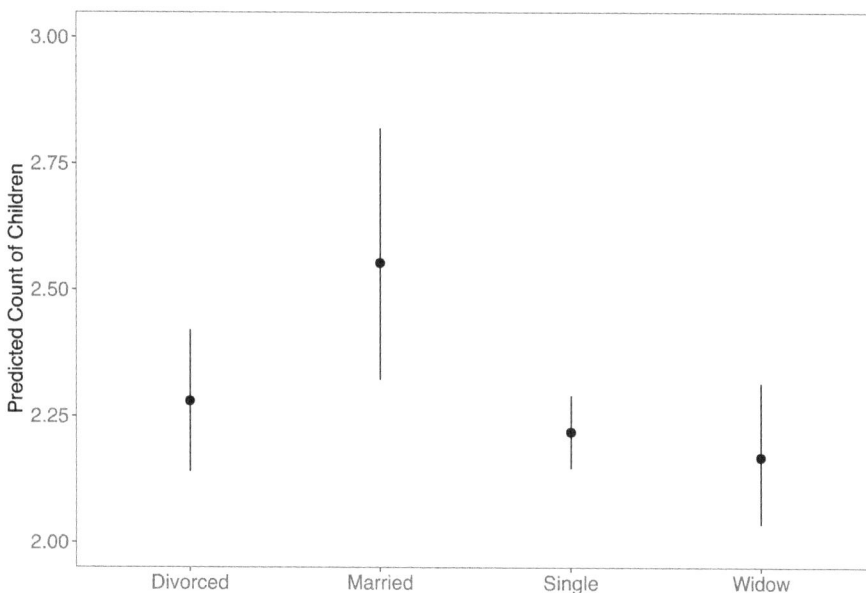

FIGURE 9.15
Predicted number of children by marital status.

Hurdle Models

We end by discussing hurdle models, which combine elements of zero-trun-cated models and zero-inflated models. Like zero-inflated models, hurdle models combine a BRM with a count model, PRM or NBRM. Unlike zero-inflated models, hurdle models do not treat the zeros as a mixture of two separate data-generating processes; instead, they treat them as two distinct processes without mixing. As a result, hurdle models allow for zero deflation as well as zero inflation (Feng 2021).

Estimation and Interpretation

The hurdle model can be estimated using the `hurdle` function from the `pscl` package. Again, we can specify a `dist="poisson"` or `dist="negbin"` option depending on whether there is overdispersion.

```
m8 <- hurdle(num.children ~ religious + minority  + female +
age + education + divorced + married + widow , dist =
"poisson", data = X)

m9 <- hurdle(num.children ~ religious + minority  + female +
age + education + divorced + married + widow , dist =
"negbin", data = X)
```

```
summary(m9)

Call:
hurdle(formula = num.children ~ religious + minority + female
+ age + education + divorced + married +
    widow, data = X, dist = "negbin")

Pearson residuals:
    Min       1Q  Median       3Q      Max
-1.8690  -0.7148  -0.0681   0.5449   5.8861

Count model coefficients (truncated negbin with log link):
              Estimate Std. Error z value Pr(>|z|)
(Intercept)   0.272151   0.132974   2.047  0.04069 *
religious     0.006699   0.007083   0.946  0.34427
minority      0.219442   0.076203   2.880  0.00398 **
female        0.054256   0.043186   1.256  0.20899
age           0.008924   0.001486   6.007 1.89e-09 ***
education    -0.016045   0.005870  -2.733  0.00627 **
divorced      0.040158   0.059048   0.680  0.49644
married       0.199473   0.069948   2.852  0.00435 **
widow        -0.034927   0.069513  -0.502  0.61534
Log(theta)   16.423042        NaN     NaN      NaN
Zero hurdle model coefficients (binomial with logit link):
              Estimate Std. Error z value Pr(>|z|)
(Intercept)  -1.339561   0.284680  -4.706 2.53e-06 ***
religious     0.036019   0.018119   1.988 0.046815 *
minority      0.047754   0.194424   0.246 0.805977
female        0.692348   0.105631   6.554 5.59e-11 ***
age           0.042303   0.003463  12.215  < 2e-16 ***
education    -0.032153   0.014817  -2.170 0.030009 *
divorced      0.715372   0.200596   3.566 0.000362 ***
married       0.701497   0.246648   2.844 0.004453 **
widow        -0.106448   0.240804  -0.442 0.658452
---
Signif. codes:  0 '***' 0.001 '**' 0.01 '*' 0.05 '.' 0.1 ' ' 1

Theta: count = 13565518.981
Number of iterations in BFGS optimization: 44
Log-likelihood: -3296 on 19 Df
```

The large θ suggests that a hurdle Poisson regression is preferred in this case, so we interpret based on the hurdle Poisson regression results from m8 instead of m9. Similar to a zero-inflated model, we can use list.coef to calculate the IRR and OR of the two portions of the model, respectively. Figure 9.16 includes a plot of these estimates.

Unlike the output for the zero-inflated models, however, the binary part of the hurdle regression is based on clearing the hurdle of having any children versus having zero children. Therefore, we interpret the coefficients and odds ratios from this portion of the model comparing the odds of *having*

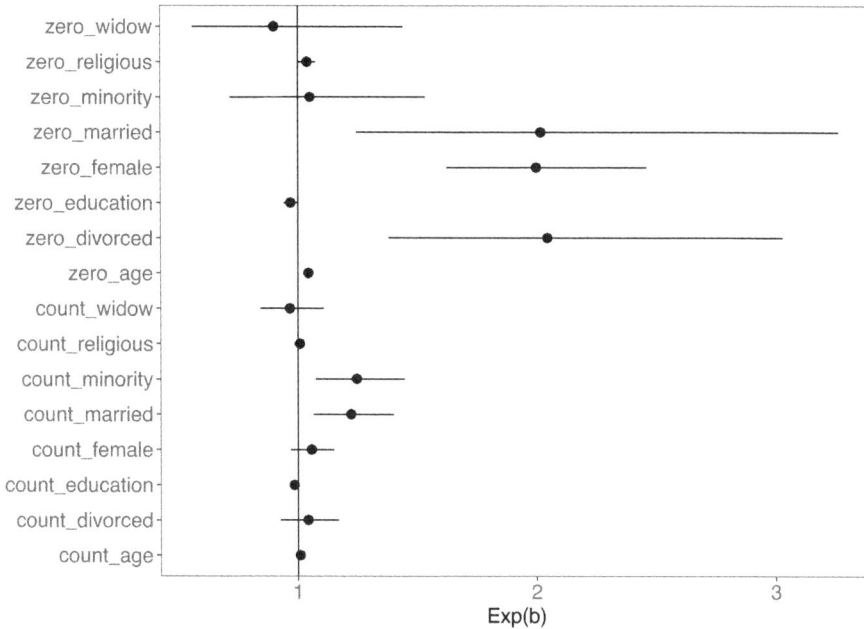

FIGURE 9.16
Exponentiated coefficients from hurdle Poisson regression.

children to the not having children rather than the odds of having zero children. Based on the hurdle regression, we can see that being married, female, and divorced is associated with about double the odds of clearing the hurdle of having children versus not. Likewise, higher levels of religiosity, lower levels of education, and being older are also associated with having higher odds of having children. Once that hurdle is cleared, being a racial or ethnic minority, married, having lower levels of education, and being older are associated with having a higher predicted number of children.

As we have done with previous models, we can use margins.des and margins.dat to generate and plot the predicted counts of children to illustrate the effect of an independent variable (Figure 9.17):

```
design <- margins.des(m8,ivs=expand.grid(education=10:18))
pdat <- margins.dat(m8,design)
ggplot(pdat,aes(x=education,y=fitted,ymin=ll,ymax=ul)) +
  theme_bw() + geom_point() + geom_line() + geom_
ribbon(alpha=.2) +
  labs(x="Education",y="Predicted # of Children")
```

Finally, we can generate the predicted probability of each level of the outcome to compare with our observed data. As shown in the resulting Figure 9.18, the hurdle model does the best job of predicting our zeros relative to the data but overcounts the number of respondents reporting one child and undercounts the number of respondents reporting two children.

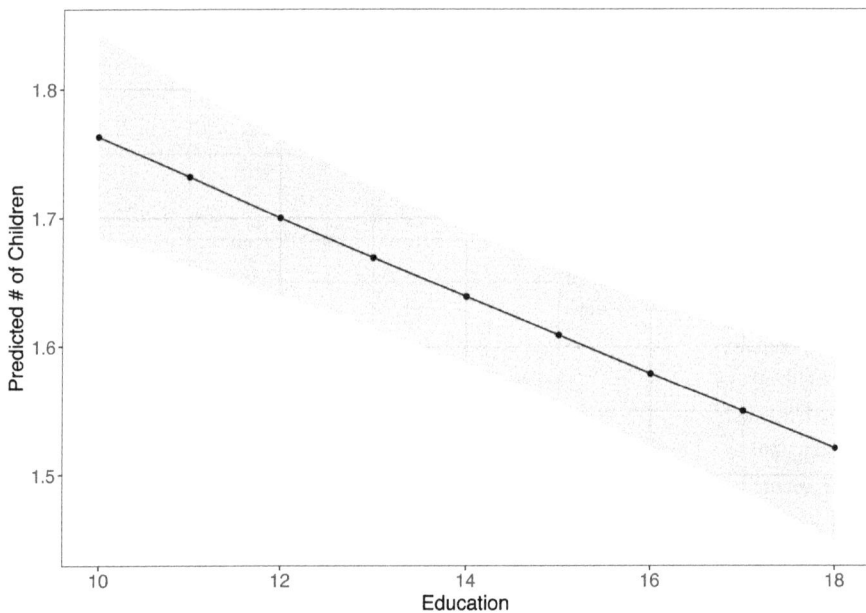

FIGURE 9.17
Predicted number of children by education.

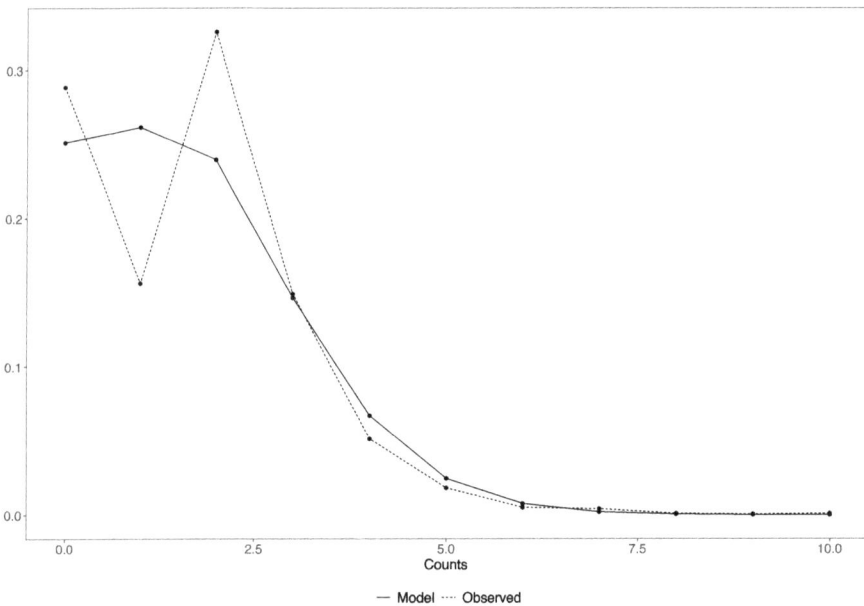

FIGURE 9.18
Predicted probabilities compared to observed proportion.

```
pdat <- data.frame(Counts=rep(0:10,2), vals=c(predict(m8,newda
ta=design,type="prob"),table(X$num.children)/length(X$num.
children)),
                     type=rep(c("model","observed"),each=11))
ggplot(pdat,aes(x=Counts,y=vals,group=type,linetype=type)) +
  theme_bw() + geom_line() + geom_point()
```

Note

1. This is a useful resource for working with quasi-likelihood methods: https://cran.r-project.org/web/packages/bbmle/vignettes/quasi.p^df

10

Additional Outcome Types

In this chapter, we consider two additional types of regression models for limited, dependent variables. First, we consider conditional or fixed effects logistic regression (Allison 2009; McFadden 1973). These models are appropriate when respondents must make a single choice from a nominal set of possible categories, such as the type of transportation they take to get somewhere (e.g., car, bus, or plane) or who is selected from a pool of people, for example, for interviews or job offers. In this context, characteristics of the alternatives are used to predict choices, which is why these are sometimes called "alternative-specific" data (e.g., Long and Freese 2006). Though respondent-level variation can be exploited too (Allison 2009a). Next, we consider the rank-ordered or exploded logistic regression model (Allison and Christakis 1994). This is a generalization of the conditional logistic regression model that is appropriate when respondents rank each response category from most to least desirable.

Below, we provide an overview of the statistical models, along with applied examples in R. We include two examples of conditional logistic regression, with different aims, and one example of rank-ordered logistic regression. The material presented in this chapter serves as an introduction to these models. More in depth treatments are available elsewhere (Allison 2009; Allison and Christakis 1994; Collett 2002; Greene 2003; McFadden 1973).

Conditional or Fixed Effects Logistic Regression

A standard conditional logistic regression model predicts the choice the respondent made as a function of attributes of the choices. This is in contrast to a standard logistic regression model discussed earlier where the choice is a function of the characteristics of the respondent. In the context of a hiring decision, for example, we might model whether the job candidate was chosen as a function of the candidate's gender, age, race, experience, and so on. In our second example, we consider whether the effect of these alternative-specific predictors varies by attributes of the respondent.

Using the same notation as we have used for other logistic regression models, the conditional logistic regression model can be written as (Allison 2009; Long 1997):

$$Pr\left(x_i = c\right) = \frac{\exp\left(X_i b\right)}{\sum_{j=1}^{J} \exp\left(X_i b\right)} \text{ for } c = 1 \text{ to } J. \tag{10.1}$$

The model says that the probability respondent *i* selects choice *c* out of *J* possible choices is the exponentiated linear predictor for that choice, divided by the sum of the exponentiated linear predictors for all choices.

Travel Choice

For our first example of a conditional logistic regression model, we examine choice of travel mode for various trips, as originally described in Greene and Jones (1997). These data were also analyzed in Long and Freese (2006). The outcome is the chosen mode of transportation; that is, whether the respondent took a train, a bus, or a car to get somewhere. We include travel time as a covariate. Conditional logistic regressions require a nested or long data structure. The alternatives are nested in respondents, allowing alternative-specific attributes to vary within respondents. For example, below is the first case in the data. In these data, id refers to the respondent; mode is a categorical variable tracking travel mode; train, bus, and car are dummy variables for whether the respondent took that form of transportation; time is the estimated amount of time it would take for the respective mode of travel to get the respondent to their destination; and choice is the mode of transportation that the respondent took to make their trip. Note that all our predictors – the dummy variables and time – vary within subjects.

```
> filter(LF06travel,id==1) %>% select(id,mode,train,bus,car,ti
me,choice)
  id mode train bus car time choice
1  1    1     1   1   0  406      0
2  1    2     0   1   0  452      0
3  1    3     0   0   1  180      1
```

There are several R packages that can estimate conditional logistic regression models. At the time of this writing, the survival package (Therneau 2022) has the most extensive options for estimating conditional logistic regression models. We use the Epi package (Carstensen et al. 2015) rather than the survival package because we found it easier to work with Epi objects in post-estimation. Specifying a conditional logistic regression model in the context of the Epi package is also more straightforward/consistent with how we have specified regression models in this book up to this point. For example, the code for our first model is presented below. The `clogistic` function implements the model. Then, we specify the prediction equation. The dependent variable (`choice`) is distributed as a function of three predictors (`car`, `bus`, and `time`). Then, we specify two additional options. First, `strata` is the unit of nesting, that is, the choices are nested in `ids`. And, second, if the variables in the model are not in the workspace, we tell it which `data` to use.

```
clogistic(choice ~ car + bus + time , strata=id, data=LF06travel)
```

Typically, we will define the model as an object so that we can easily use post-estimation commands on it. Below we define the conditional logistic regression model as an object, and then print the output:

```
> m1 <- clogistic(choice ~ car + bus + time , strata=id,
data=LF06travel)
> m1

Call:
clogistic(formula = choice ~ car + bus + time, strata = id,
data = LF06travel)

       coef exp(coef) se(coef)     z       p
car  -1.4951     0.224  0.29635 -5.05 4.5e-07
bus  -0.4877     0.614  0.29656 -1.64 1.0e-01
time -0.0201     0.980  0.00237 -8.46 0.0e+00

Likelihood ratio test=153   on 3 df, p=0, n=456
```

The output includes the call, the estimated regression coefficients, exponentiated regression coefficients (odds ratios), the estimated standard error for each coefficient, the z-statistic, and its p-value. The bottom of the output includes the results from an LR test, comparing the full model to a null model with no predictors.

As in logistic regression, the coefficient can be interpreted in terms of change in the log-odds of being chosen. Net of travel time, respondents are less likely to take a car or a bus than to take a train. Similarly, each minute longer of estimated travel time results in a .02 decrease in the log-odds of selecting that mode of transportation. The odds ratios have a more straightforward interpretation than the log-odds coefficients. The odds of taking a train are 4.46 times the odds of taking a car (i.e., 1/.224 = 4.46), net of travel time, and other modes of transportation. The odds of taking a train are 1.63 times the odds of taking a bus, net of travel time, and other modes of transportation. Or, respondents are 63% more likely to take a train than a bus, net of travel time, and other modes of transportation. Finally, the odds ratio for `time` can be interpreted as: for each minute shorter of travel time, respondents are 2% more (i.e., 1/.98 = 1.02) likely to choose that mode of transit.

As we have noted throughout this book, predicted values or probabilities are a preferred means to illustrate model effects. We can use the `margins.dat.clogit` function in `catregs` to generate predicted probabilities for each response alternative. The function takes a conditional logistic regression model object and a design matrix of independent variable values. By default, inference is supported via simulation. The function simulates 1,000 (by default) draws off of a multivariate normal distribution defined by the estimated regression coefficients and its variance/covariance matrix (via the MASS package; Ripley et al. 2013), and uses those values to generate a sampling distribution of predicted probabilities. The upper and lower limits in the output are the simulated values at the corresponding *alpha* value. The estimate of the standard error (`se`) is the standard deviation of the observed/simulated sampling distribution. Inference is also supported via bootstrapping, though this is not estimated by default. From the example above, below we compute the probability distribution when each mode takes 60 minutes:

```
> design <- data.frame(car=c(0,0,1),bus=c(0,1,0),tim
e=c(60,60,60))
> design
  car bus time
1   0   0   60
2   0   1   60
3   1   0   60
> ma1 <- margins.dat.clogit(m1,design)
> ma1
  car bus time    lp probs    ll    ul    se
1   0   0   60 0.300 0.544 0.423 0.653 0.059
2   0   1   60 0.184 0.334 0.219 0.462 0.061
3   1   0   60 0.067 0.122 0.071 0.193 0.031
```

The prediction matrix, `design`, includes a row for each response alternative. We included dummy variables for response alternatives in `m1`, so row 1 is for the reference category (train). As is illustrated in Figure 10.1, when it takes an hour by each mode, there is a .54 probability of taking the train, a .33 probability of taking the bus, and a .12 probability of driving. Per Eq. 10.1, `lp` refers to the linear predictions for each response alternative, and the probabilities simply constrain the linear predictions to sum to one. The `ll` and `ul` columns refer to the lower and upper limit, respectively, of the inner 95% of the simulated sampling distribution. These limits define the bounds of confidence intervals in Figure 10.1.

In this case, it is quite clear that respondents are significantly more likely to take the train than to ride the bus or drive, and they are significantly more likely take a bus than to drive. This is apparent in that none of the confidence intervals overlap. We can make direct comparisons using `compare.margins` too. Below is the code to explicitly test whether the predicted probabilities vary by mode, with code for each pairwise comparison.

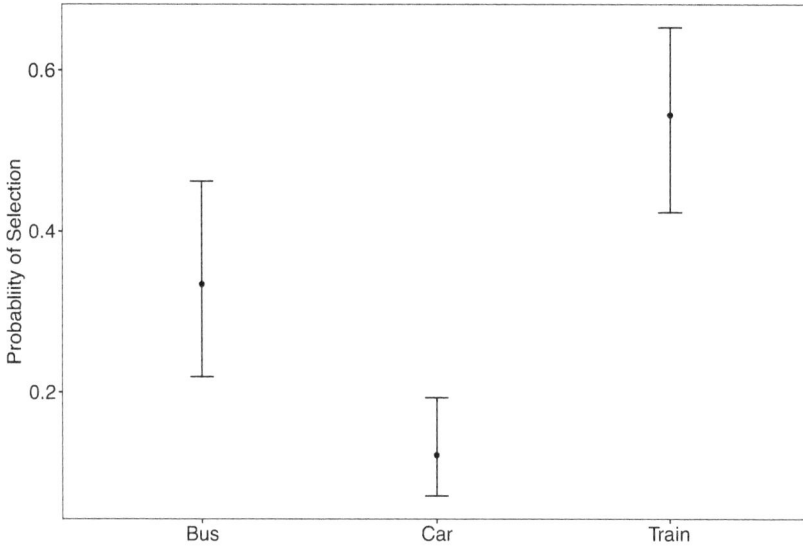

FIGURE 10.1
Predicted probabilities of selecting a mode of transportation given that each would take 60 minutes.

```
> compare.margins(margins=c(ma1$probs)[1:2],margins.
ses=c(ma1$se)[1:2])
$difference
[1] 0.21

$p.value
[1] 0.006

> compare.margins(margins=c(ma1$probs)[c(1,3)],margins.
ses=c(ma1$se)[c(1,3)])
$difference
[1] 0.422

$p.value
[1] 0

> compare.margins(margins=c(ma1$probs)[2:3],margins.
ses=c(ma1$se)[2:3])
$difference
[1] 0.212

$p.value
[1] 0.001
```

The probability of taking the train is .54 and the probability of taking the bus is .33. As noted in the output above, the difference in these probabilities is .21, and the probability that we observed a difference that large if the difference in the population were zero is.006, that is, $p =.006$. Similarly the difference in probabilities of .422 between taking the train and driving is significant ($p <.001$), and the difference in probabilities of .212 between taking the bus and driving is significant ($p =.001$).

Social scientists are often interested in how case or unit-level variables moderate the effect of alternative-specific effects. Unfortunately, in the context of fixed effects models, we cannot estimate coefficients associated with variables that only have variation between clusters or units (Allison 2009). Briefly this is due to adjustments that are made to the data during estimation, which remove between-group or between-cluster variation. We can estimate, however, the extent to which the effects of within-cluster variables vary by between-cluster variables. That is, how cluster-level variables moderate alternative-specific effects. For example, in selecting an applicant to hire, the effect of the applicant's gender may vary by the education of the person making the decision. We illustrate such a model with our next example.

Occupational Preferences

Logan (1983) introduced a model of mobility processes based on the conditional logistic regression. To illustrate the model, he used data from the General Social Survey (GSS) from 1972 to 1978. We provide the code in the online supplement to apply conditional logistic regression models to those data. Here, we apply conditional logistic regression models to mobility processes in the 2016 GSS. We analyze data from the 2,083 respondents with complete data on respondent and father's social class position, education, and race. We first examine the distribution of respondents in each alternative, which is as follows:

```
 Unskilled Manual   Skilled Manual   Self-Employed Nonmanual/Service
           0.156            0.114               0.118            0.167
 Professional, Lower Professional, Higher
           0.256                   0.191
```

This distribution should form the basis of a probability distribution if we apply a conditional logistic regression model to the occupational category for the respondent (Logan 1983). Below, we fit such a conditional logistic regression including only alternative-specific dummy variables, generate a design matrix to generate predictions, and then estimate predicted probabilities.

```
> m2 <- clogistic(occ ~ skmanual+ selfemp + service + proflow +
profhigh, strata=id, data=gss2016)

> design1 <- data.frame(skmanual=c(0,1,0,0,0,0),selfemp=c(0,0,1,0,0,0)
,service=c(0,0,0,1,0,0),proflow=c(0,0,0,0,1,0),profhig
h=c(0,0,0,0,0,1))

> margins.dat.clogit(m2,design1)
```

	skmanual	selfemp	service	proflow	profhigh	lp	probs	ll	ul	se
1	0	0	0	0	0	1.000	0.156	0.141	0.171	0.008
2	1	0	0	0	0	0.731	0.114	0.101	0.128	0.007
3	0	1	0	0	0	0.756	0.118	0.105	0.131	0.007
4	0	0	1	0	0	1.071	0.167	0.151	0.183	0.008
5	0	0	0	1	0	1.645	0.256	0.239	0.276	0.010
6	0	0	0	0	1	1.225	0.191	0.174	0.208	0.008

As illustrated above, the first row of design1 is all zeros, which will estimate the probability of the reference category (unskilled manual in this case).
Note that probs matches the observed distribution of occupational categories. Logan's (1983) insight is that we can include covariates to examine how
individual-level predictors modify the probability of being in specific occupational categories. Our next model includes paternal occupational category
as a factor, using the same occupational categories.

Defining the statistical model itself is straightforward: we interact the
dummy variables for occupational categories with a factor variable for
paternal education (pclass). Below is the model specification and corresponding output. Note that the main effects of pclass are omitted (NA) – it
does not vary within units and cannot be estimated, but the other terms
in the model are estimated. Without the fixed effects being estimated, the
interaction terms tell us how the main effect of each dummy variable is
conditioned by paternal occupational categories. In the output, unskilled
manual labor is the reference category for respondent occupations and
unskilled manual labor is also the reference category for father's occupation. So, the main effect for skilled manual labor of −.546 means that
respondents who had fathers who were unskilled manual workers are
less likely to be skilled manual workers than unskilled workers. The first
interaction term (skmanual:pclassSkilled Manual) of .723 means that
respondents who had fathers who were skilled manual workers are more
likely to be skilled manual workers than respondents who had fathers who
were unskilled workers. Interpreting such interaction terms is not straightforward, especially when there are several alternatives. Accordingly, we
recommend examining predicted probabilities.

```
> m3 <- clogistic(occ ~ skmanual*pclass + selfemp*pclass + service*pclass +
proflow*pclass + profhigh*pclass, strata=id, data=gss2016)
> m3

Call:
clogistic(formula = occ ~ skmanual * pclass + selfemp * pclass +
    service * pclass + proflow * pclass + profhigh * pclass,
    strata = id, data = gss2016)
```

	coef	exp(coef)	se(coef)	z	p
skmanual	-0.5457	0.579	0.160	-3.4190	6.3e-04
pclassSkilled Manual	NA	NA	0.000	NA	NA
pclassSelf-Employed	NA	NA	0.000	NA	NA
pclassNon-Manual/Service	NA	NA	0.000	NA	NA

```
pclassProfessional, Lower            NA      NA   0.000     NA      NA
pclassProfessional, Higher           NA      NA   0.000     NA      NA
selfemp                         -0.8227   0.439   0.175 -4.7012 2.6e-06
service                         -0.0189   0.981   0.137 -0.1374 8.9e-01
proflow                          0.0457   1.047   0.135  0.3378 7.4e-01
profhigh                        -0.5457   0.579   0.160 -3.4190 6.3e-04
skmanual:pclassSkilled Manual    0.7234   2.061   0.226  3.2034 1.4e-03
skmanual:pclassSelf-Employed    -0.0646   0.937   0.246 -0.2624 7.9e-01
skmanual:pclassNon-Manual/Service 0.6635  1.942   0.511  1.2972 1.9e-01
skmanual:pclassProfessional, Lower -0.1475 0.863  0.438 -0.3364 7.4e-01
skmanual:pclassProfessional, Higher 0.2039 1.226  0.298  0.6843 4.9e-01
pclassSkilled Manual:selfemp     0.0724   1.075   0.272  0.2662 7.9e-01
pclassSelf-Employed:selfemp      0.9056   2.473   0.233  3.8850 1.0e-04
pclassNon-Manual/Service:selfemp 0.1295   1.138   0.637  0.2034 8.4e-01
pclassProfessional, Lower:selfemp 0.9768  2.656   0.366  2.6705 7.6e-03
pclassProfessional, Higher:selfemp 1.1169 3.055   0.277  4.0370 5.4e-05
pclassSkilled Manual:service    -0.0093   0.991   0.217 -0.0429 9.7e-01
pclassSelf-Employed:service      0.0189   1.019   0.209  0.0904 9.3e-01
pclassNon-Manual/Service:service 0.4243   1.529   0.477  0.8902 3.7e-01
pclassProfessional, Lower:service 0.6838  1.981   0.321  2.1308 3.3e-02
pclassProfessional, Higher:service 0.1655 1.180   0.261  0.6350 5.3e-01
pclassSkilled Manual:proflow     0.2928   1.340   0.205  1.4275 1.5e-01
pclassSelf-Employed:proflow      0.3802   1.463   0.197  1.9329 5.3e-02
pclassNon-Manual/Service:proflow 0.2728   1.314   0.484  0.5637 5.7e-01
pclassProfessional, Lower:proflow 1.3406  3.821   0.296  4.5265 6.0e-06
pclassProfessional, Higher:proflow 1.0441 2.841   0.231  4.5167 6.3e-06
pclassSkilled Manual:profhigh    0.6511   1.918   0.228  2.8588 4.3e-03
pclassSelf-Employed:profhigh     0.6056   1.832   0.222  2.7234 6.5e-03
pclassNon-Manual/Service:profhigh 0.0757  1.079   0.592  0.1279 9.0e-01
pclassProfessional, Lower:profhigh 1.6984 5.465   0.314  5.4094 6.3e-08
pclassProfessional, Higher:profhigh 1.5809 4.860  0.247  6.3939 1.6e-10

Likelihood ratio test=326  on 30 df, p=0, n=12498
```

To generate predicted probabilities from m3, we need to define the design matrix. It is important to generate the design matrix so that the columns correspond to the estimated terms in the conditional logistic regression model. The order of terms is printed in the output above. In generating the design matrix, we just ignore the missing coefficients. Margins.dat.clogit will automatically remove missing coefficients and adjust the variance/covariance matrix when generating predicted probabilities.

We start by defining the dummy variables for the response categories, in the order of the estimated parameters. Next, we add dummy variables for the interactions. The naming convention we adopted for the dummy variables is that the first two letters denote the respondent's occupation and the second set of two letters following the period denote the respondent's paternal occupation. We abbreviated the occupational categories as follows: sk=skilled manual, se=self-employed, sr=service, lp=lower professional, and up=upper professional. In this case, we will estimate the probability distribution for respondents with fathers who were upper professionals. This is indicated by the 1s in the interactions with upper-level professional fathers (i.e., ##.up=1).

```
> design1 <- data.frame(skmanual=c(0,1,0,0,0,0),selfemp=c(0,0,1,0,0,0),service=c(0,0,0,1,0,0),profl
ow=c(0,0,0,0,1,0),profhigh=c(0,0,0,0,0,1))
> design1 <- mutate(design1,
+     sk.sk=skmanual*0,sk.se=skmanual*0,sk.sr=skmanual*0,sk.lp=skmanual*0,sk.up=skmanual*1,
+     se.sk=selfemp*0,se.se=selfemp*0,se.sr=selfemp*0,se.lp=selfemp*0,se.up=selfemp*1,
+     sr.sk=service*0,sr.se=service*0,sr.sr=service*0,sr.lp=service*0,sr.up=service*1,
+     lp.sk=proflow*0,lp.se=proflow*0,lp.sr=proflow*0,lp.lp=proflow*0,lp.up=proflow*1,
+     up.sk=service*0,up.se=profhigh*0,up.sr=profhigh*0,up.lp=profhigh*0,up.up=profhigh*1)
>
> mar6<-margins.dat.clogit(m3,design1,rounded=3)
> mar6
```

	skmanual	selfemp	service	proflow	profhigh	sk.sk	sk.se	sk.sr	sk.lp	sk.up	se.sk	se.se	se.sr	se.lp
1	0	0	0	0	0	0	0	0	0	0	0	0	0	0
2	1	0	0	0	0	0	0	0	0	1	0	0	0	0
3	0	1	0	0	0	0	0	0	0	0	0	0	0	0
4	0	0	1	0	0	0	0	0	0	0	0	0	0	0
5	0	0	0	1	0	0	0	0	0	0	0	0	0	0
6	0	0	0	0	1	0	0	0	0	0	0	0	0	0

	se.up	sr.sk	sr.se	sr.sr	sr.lp	sr.up	lp.sk	lp.se	lp.sr	lp.lp	lp.up	up.sk	up.se	up.sr	up.lp	up.up
1	0	0	0	0	0	0	0	0	0	0	0	0	0	0	0	0
2	0	0	0	0	0	0	0	0	0	0	0	0	0	0	0	0
3	1	0	0	0	0	0	0	0	0	0	0	0	0	0	0	0
4	0	0	0	0	0	1	0	0	0	0	0	0	0	0	0	0
5	0	0	0	0	0	0	0	0	0	0	1	0	0	0	0	0
6	0	0	0	0	0	0	0	0	0	0	0	0	0	0	0	1

```
      lp probs    ll     ul     se
1 1.000 0.100 0.073 0.131 0.015
2 0.711 0.071 0.048 0.100 0.013
3 1.342 0.134 0.103 0.173 0.018
4 1.158 0.116 0.087 0.153 0.017
5 2.974 0.297 0.251 0.343 0.023
6 2.816 0.282 0.232 0.327 0.024
```

The probability distribution above is the top row of Figure 10.2. In the sup-porting materials, we used m3 to generate predicted probabilities for each level of paternal occupation. Figure 10.2 plots those predicted probabilities, illustrating the implications of paternal occupation on respondent occupation. For example, respondents with fathers who had a higher-level professional occupation had a .282 probability of becoming a higher-level professional, whereas respondents with a father who was an unskilled manual worker had a .125 probability of becoming an upper-level professional. The differ-ence between these probabilities, as estimated via compare.margins, is sig-nificant ($p < .001$), and it illustrates the implications of paternal occupational category on respondent occupational category.

Unlike the log-linear model that we discussed in Chapter 3, the conditional logistic regression model allows continuous control variables (Logan 1983). To that end, we included the respondent's education in another conditional logis-tic regression model. This adds five terms to the model – one for each response alternative. Below are the model output and the code to generate predicted probabilities for respondents with fathers who were unskilled manual work-ers, and who achieved 12 years of schooling. Education is set by the 12s in the interactions with education (i.e., ##.e=12).

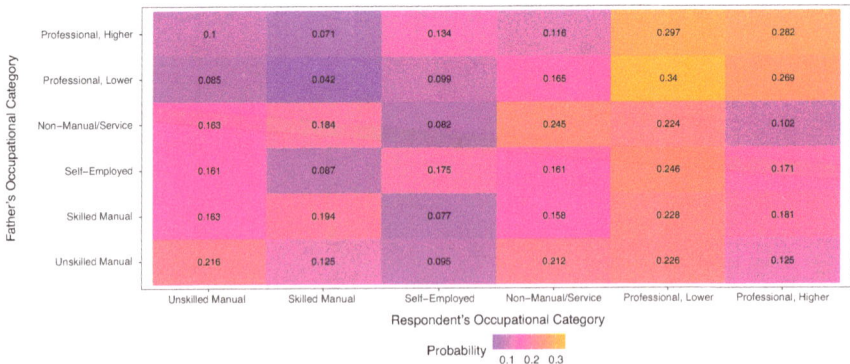

FIGURE 10.2
Heatmap of probabilities illustrating how respondent occupational category is affected by paternal occupational category.

```
> m4 <- clogistic(occ ~ skmanual*pclass + selfemp*pclass + service*pclass + proflow*pclass +
profhigh*pclass + skmanual*educ + selfemp*educ + service*educ + proflow*educ + profhigh*educ,
strata=id, data=gss2016)
> m4
```

Call:
```
clogistic(formula = occ ~ skmanual * pclass + selfemp * pclass +
    service * pclass + proflow * pclass + profhigh * pclass +
    skmanual * educ + selfemp * educ + service * educ + proflow *
    educ + profhigh * educ, strata = id, data = gss2016)
```

	coef	exp(coef)	se(coef)	z	p
skmanual	-0.8044	0.447349	0.4316	-1.8639	6.2e-02
pclassSkilled Manual	NA	NA	0.0000	NA	NA
pclassSelf-Employed	NA	NA	0.0000	NA	NA
pclassNon-Manual/Service	NA	NA	0.0000	NA	NA
pclassProfessional, Lower	NA	NA	0.0000	NA	NA
pclassProfessional, Higher	NA	NA	0.0000	NA	NA
selfemp	-4.4265	0.011956	0.4950	-8.9430	0.0e+00
service	-2.4621	0.085255	0.4283	-5.7488	9.0e-09
proflow	-6.8138	0.001099	0.4588	-14.8508	0.0e+00
profhigh	-7.5992	0.000501	0.4960	-15.3211	0.0e+00
educ	NA	NA	0.0000	NA	NA
skmanual:pclassSkilled Manual	0.7159	2.046011	0.2261	3.1658	1.5e-03
skmanual:pclassSelf-Employed	-0.0601	0.941652	0.2462	-0.2442	8.1e-01
skmanual:pclassNon-Manual/Service	0.6453	1.906567	0.5123	1.2597	2.1e-01
skmanual:pclassProfessional, Lower	-0.1744	0.839930	0.4404	-0.3961	6.9e-01
skmanual:pclassProfessional, Higher	0.1765	1.193044	0.3011	0.5862	5.6e-01
pclassSkilled Manual:selfemp	-0.0326	0.967927	0.2756	-0.1183	9.1e-01
pclassSelf-Employed:selfemp	0.8240	2.279574	0.2387	3.4514	5.6e-04

```
pclassNon-Manual/Service:selfemp     -0.1125   0.893580   0.6454   -0.1743   8.6e-01
pclassProfessional, Lower:selfemp     0.6021   1.825901   0.3735    1.6119   1.1e-01
pclassProfessional, Higher:selfemp    0.7163   2.046937   0.2853    2.5109   1.2e-02
pclassSkilled Manual:service         -0.0813   0.921885   0.2194   -0.3706   7.1e-01
pclassSelf-Employed:service          -0.0121   0.987935   0.2124   -0.0572   9.5e-01
pclassNon-Manual/Service:service      0.2612   1.298511   0.4822    0.5417   5.9e-01
pclassProfessional, Lower:service     0.4317   1.539801   0.3263    1.3229   1.9e-01
pclassProfessional, Higher:service   -0.1017   0.903304   0.2669   -0.3811   7.0e-01
pclassSkilled Manual:proflow          0.0984   1.103404   0.2200    0.4472   6.5e-01
pclassSelf-Employed:proflow           0.1043   1.109973   0.2148    0.4858   6.3e-01
pclassNon-Manual/Service:proflow     -0.2170   0.804909   0.5225   -0.4154   6.8e-01
pclassProfessional, Lower:proflow     0.6189   1.856793   0.3165    1.9553   5.1e-02
pclassProfessional, Higher:proflow    0.2734   1.314425   0.2506    1.0908   2.8e-01
pclassSkilled Manual:profhigh         0.4513   1.570405   0.2420    1.8653   6.2e-02
pclassSelf-Employed:profhigh          0.3170   1.372988   0.2394    1.3243   1.9e-01
pclassNon-Manual/Service:profhigh    -0.4299   0.650579   0.6260   -0.6867   4.9e-01
pclassProfessional, Lower:profhigh    0.9567   2.603130   0.3341    2.8639   4.2e-03
pclassProfessional, Higher:profhigh   0.7895   2.202284   0.2661    2.9674   3.0e-03
skmanual:educ                         0.0219   1.022127   0.0339    0.6460   5.2e-01
selfemp:educ                          0.2888   1.334781   0.0366    7.8970   2.9e-15
service:educ                          0.1993   1.220528   0.0330    6.0387   1.6e-09
proflow:educ                          0.5237   1.688342   0.0336   15.5646   0.0e+00
profhigh:educ                         0.5370   1.710940   0.0354   15.1650   0.0e+00

Likelihood ratio test=828   on 35 df, p=0, n=12498
>
> # Margins when education is 12
> design1 <- data.frame(skmanual=c(0,1,0,0,0,0),selfemp=c(0,0,1,0,0,0),service=c(0,0,1,0,0),profl
ow=c(0,0,0,1,0),profhigh=c(0,0,0,0,1),
> design1 <- mutate(design1,
```

```
+            sk.sk=skmanual*0,sk.se=skmanual*0,sk.sr=skmanual*0,sk.lp=skmanual*0,sk.
up=skmanual*1,
+            se.sk=selfemp*0,se.se=selfemp*0,se.sr=selfemp*0,se.lp=selfemp*0,se.
up=selfemp*1,
+            sr.sk=service*0,sr.se=service*0,sr.sr=service*0,sr.lp=service*0,sr.
up=service*1,
+            lp.sk=proflow*0,lp.se=proflow*0,lp.sr=proflow*0,lp.lp=proflow*0,lp.
up=proflow*1,
+            up.sk=profhigh*0,up.se=profhigh*0,up.sr=profhigh*0,up.lp=profhigh*0,up.
up=profhigh*1,
+            sk.e=skmanual*12,
+            se.e=selfemp*12,
+            sr.e=service*12,
+            lp.e=proflow*12,
+            up.e=profhigh*12)
> margins.dat.clogit(m4,design1)
```

	skmanual	selfemp	service	proflow	profhigh	sk.sk	sk.se	sk.sr	sk.lp	sk.up	up.sk	up.se	up.sr	up.lp	up.up	sk.e	se.e	sr.e	lp.e	up.e	lp	probs
1	0	0	0	0	0	0	0	0	0	0	0	0	0	0	0	0	0	0	0	0	1.000	0.209
2	0	0	0	0	0	1	0	0	0	0	0	0	0	0	0	12	0	0	0	0	0.694	0.145
3	0	0	0	0	0	0	0	0	0	0	0	12	0	0	0	0	12	0	0	0	0.783	0.164
4	0	0	0	0	0	0	0	0	1	0	0	0	0	0	0	0	0	12	0	0	0.842	0.176
5	0	0	0	0	0	0	0	0	0	0	0	0	0	12	0	0	0	0	12	0	0.775	0.162
6	0	0	0	0	1	0	0	0	0	1	0	0	0	0	0	1	0	0	0	12	0.694	0.145

	sr.sr	sr.lp	sr.up	lp.sk	lp.se	lp.sr	lp.lp	lp.up	se.up	sr.sk	sr.se
1	0	0	0	0	0	0	0	0	0	0	0
2	1	0	0	0	0	0	0	0	0	0	1
3	0	0	0	0	0	0	0	0	1	0	0
4	0	1	0	0	0	0	1	0	0	0	0
5	0	0	0	0	0	0	0	0	0	0	0
6	0	0	0	0	0	0	0	0	0	0	0

```
         ll     ul     se
1 0.155 0.274 0.030
2 0.101 0.200 0.026
3 0.121 0.213 0.024
4 0.133 0.226 0.025
5 0.126 0.202 0.020
6 0.111 0.185 0.019
```

The probability distribution above is again the top row of Figure 10.3. The figure shows how respondent education modifies the effect of paternal occupational category. For example, respondents with 12 years of schooling and fathers with an unskilled manual occupation had a .083 probability of obtaining a higher-level professional occupation. Respondents with 16 years of schooling and fathers with an unskilled manual occupational had a .218 probability of obtaining a higher-level professional occupation, a difference of .135 ($p<.001$). Education plays less of a role in determining who becomes self-employed. Consider respondents who had self-employed fathers. At 12 years of schooling, the probability respondents are self-employed is .197, and at 16 years of schooling, the probability respondents are self-employed is .179, a difference of .018 ($p=.271$).

As this example illustrates, conditional logistic regression models can be informative for understanding alternative-specific choices. In line with most social science applications of conditional logistic regression models, this example also illustrates how to include individual-level covariates that only vary between individuals. Below, we discuss a generalized version of the conditional logistic regression model tailored to data where choices are rank-ordered.

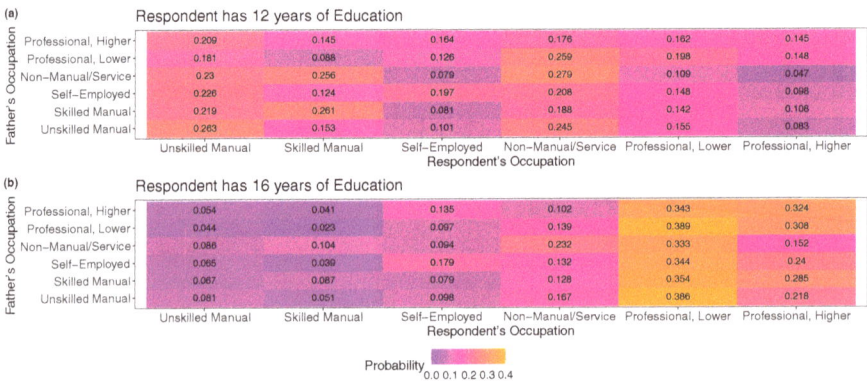

FIGURE 10.3
Heatmap of probabilities illustrating how respondent occupational category is affected by paternal occupational category after controlling for respondent education.

Rank-Ordered or Exploded Logistic Regression

The conditional logistic regression model is appropriate when respondents make one selection from a set of alternative-specific choices. A similar data structure occurs when respondents are asked to rank order all possible choices from a set of alternative-specific choices. This leads to the rank-ordered or exploded logistic regression model (Allison and Christakis 1994). It is sometimes called an "exploded logistic regression" because the use of the logit formula is expanded or "exploded" to adjust for multiple choices being made (Long and Freese 2006). Consider the first choice being made – it is effectively a single conditional logistic regression model. The model for the first choice/rank can defined as:

$$Pr\left(x_1 = m_1 \mid X_i\right) = \frac{\exp\left(X_i b_{m_{1|b}}\right)}{\sum_{j=1}^{J} \exp\left(X_i b_{m_{j|b}}\right)} \text{ for } m=1 \text{ to } J. \qquad (10.2)$$

where $x_1 = m$ indicates that alternative m is the first choice, $Pr\left(x_1 = m_1 \mid X_i\right)$ refers to the probability of the first choice, X_i contains case-specific variables or attributes, b is the base alternative, and as in conditional logistic regression, and $b_{m_{1|b}}$ refers to the effect of each variable on the log-odds of choosing alternative m over the base alternative. As noted, Eq. 10.2 defines the model for the first choice only. Conditional on the first choice, participants make a second choice, and so on. The second choice can be defined as follows:

$$Pr\left(x_2 = m_2 \mid X_i, x_1\right) = \frac{\exp\left(X_i b_{m_{2|b}}\right)}{\left[\sum_{j=1}^{J} \exp\left(X_i b_{m_{j|b}}\right)\right] - \exp\left(X_i b_{m_{1|b}}\right)} \text{ for } m=1 \text{ to } J. \qquad (10.3)$$

Notice that here we are modeling the respondent's second rank-ordered preference and that the denominator adjusts for the probability of the respondents' first rank-ordered preference (via $\exp\left(X_i b_{m_{1|b}}\right)$). This process is repeated until there are two choices left to assign probabilities. The repeated use of the logit link to estimate the conditional choices is why this model is sometimes called the "exploded logit" model.

As an example, we apply this model to video game rankings (Croissant 2012). Respondents were asked to rank each of the following video game consoles: Gameboy, GameCube, Personal Computer, Playstation, Playstation Portable, and Xbox. We model respondent preferences as a function of whether they own each of the consoles, the number of hours they spend gaming per week, and their age. As in conditional logistic regression, the data structure for a rank-ordered logistic regression model should be nested.

Specifically, each alternative is nested in respondents, so each respondent should have six rows, corresponding to each outcome category. Below is the code to estimate the model. In the mlogit statement, alternative-specific predictors precede the vertical dash and respondent-level predictors follow the vertical dash. We have chosen to set "PC" as the reference category.

```
> m1<- mlogit(ch ~ own | hours + age, G, reflevel = "PC")
> summary(m1)

Call:
mlogit(formula = ch ~ own | hours + age, data = G, reflevel = "PC",
    method = "nr")

Frequencies of alternatives:choice
        PC      GameBoy    GameCube PlayStation  PSPortable        Xbox
   0.17363     0.13846     0.13407     0.18462     0.17363     0.19560

nr method
5 iterations, 0h:0m:0s
g'(-H)^-1g = 6.74E-06
successive function values within tolerance limits

Coefficients :
                          Estimate Std. Error z-value  Pr(>|z|)
(Intercept):GameBoy       1.570379   1.600251  0.9813 0.3264288
(Intercept):GameCube      1.404095   1.603483  0.8757 0.3812185
(Intercept):PlayStation   2.278506   1.606986  1.4179 0.1562270
(Intercept):PSPortable    2.583563   1.620778  1.5940 0.1109302
(Intercept):Xbox          2.733774   1.536098  1.7797 0.0751272 .
own                       0.963367   0.190396  5.0598 4.197e-07 ***
hours:GameBoy            -0.235611   0.052130 -4.5197 6.193e-06 ***
hours:GameCube           -0.187070   0.051021 -3.6665 0.0002459 ***
hours:PlayStation        -0.129196   0.044682 -2.8915 0.0038345 **
hours:PSPortable         -0.233688   0.049412 -4.7294 2.252e-06 ***
hours:Xbox               -0.173006   0.045698 -3.7858 0.0001532 ***
age:GameBoy              -0.073587   0.078630 -0.9359 0.3493442
age:GameCube             -0.067574   0.077631 -0.8704 0.3840547
age:PlayStation          -0.067006   0.079365 -0.8443 0.3985154
age:PSPortable           -0.088669   0.079421 -1.1164 0.2642304
age:Xbox                 -0.066659   0.075205 -0.8864 0.3754227
---
Signif. codes:  0 '***' 0.001 '**' 0.01 '*' 0.05 '.' 0.1 ' ' 1

Log-Likelihood: -516.55
McFadden R^2:  0.36299
Likelihood ratio test : chisq = 588.7 (p.value = < 2.22e-16)
```

The parameter estimates in the output are on the log-odds metric. For example, owning a gaming console results in a .96 increase in the log-odds of ranking a console higher. For hours and age, the reference effect is for PC, so each hour spent playing video games is associated with a decrease in ranking all consoles other than PC. Likewise, increases in age are associated with a decrease in ranking all consoles other than PC.

Setting up a design matrix for an exploded logistic regression is slightly more straightforward than it is for conditional logistic regression models. The alternative-specific variables may vary within and the respondent-level variables can only vary between. Below we present two examples. In the first, we show the probability distribution when the respondent does not own any gaming consoles and the second shows the probability distribution when the respondent owns a PC, a PlayStation, and an Xbox (the first, fourth, and sixth consoles in the output). Owning the consoles has a large impact, with probabilities for PC, Playstation, and Xbox increasing in the second set of margins.

```
> design <- data.frame(own=c(0,0,0,0,0,0),hours=0,age=35)
> margins.dat.clogit(m1,design)
  own hours age probs    ll    ul    se
1   0     0  35 0.210 0.032 0.555 0.142
2   0     0  35 0.077 0.009 0.296 0.075
3   0     0  35 0.080 0.010 0.287 0.072
4   0     0  35 0.196 0.026 0.579 0.145
5   0     0  35 0.125 0.014 0.431 0.109
6   0     0  35 0.313 0.051 0.684 0.165

> design3 <- data.frame(own=c(1,0,0,1,0,1),hours=0,age=35)
> margins.dat.clogit(m1,design3)
  own hours age probs    ll    ul    se
1   1     0  35 0.254 0.039 0.634 0.162
2   0     0  35 0.035 0.004 0.164 0.044
3   0     0  35 0.037 0.004 0.172 0.042
4   1     0  35 0.237 0.031 0.644 0.164
5   0     0  35 0.058 0.006 0.246 0.066
6   1     0  35 0.379 0.061 0.751 0.186
```

As above, we can use compare.margins to assess whether specific predicted probabilities differ. Consider a second example where we compute the probabilities of selecting each alternative for the average respondent. We can illustrate the effect of hours spent playing video games on the selection probabilities by increasing the number of hours in a second set of probabilities and then comparing them. Figure 10.3 illustrates two sets of selection probabilities. In Figure 10.3a, all covariates are set to their means. In Figure 10.3b, all covariates are set to their means except for age, which is set to its mean plus one standard deviation. We used compare.margins to compute the p-value associated with how the predicted probabilities change as a result of respondents being a standard deviation older in Figure 10.4. We find that older respondents are more likely to select a computer (est.$=.2$, $p=.007$), less likely to select a Gameboy (est.$=-.026$, $p=.034$), and less likely to select PS Portable (est. $-.05$, $p=.017$).

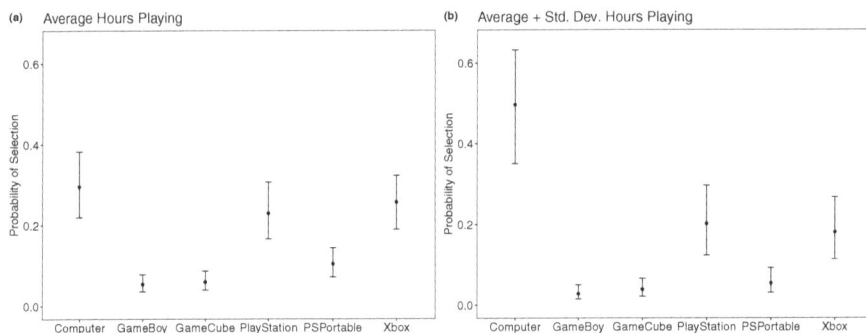

FIGURE 10.4
Probabilities of selection from a rank-ordered logistic regression model.

In the context of rank-ordered models, it is somewhat common for researchers to allow respondents to list ties. That is, two or more alternative-specific outcome states that receive the same ranking. There are multiple methods for dealing with ties in the estimation process (Allison and Christakis 1994) but none that are implemented in R at the time of this writing. Stata, on the other hand, has an impressive suite of functions for modeling alternative-specific outcomes.

Summary

In this chapter, we reviewed two classes of models for alternative-specific outcomes – the conditional logistic regression model and the rank-ordered or exploded logistic regression model. These models are both estimated on "long" data that entail a nested data structure (Wickham and Grolemund 2016). In this way, they are distinct from the models discussed earlier in this book. Practically, that means that we did not rely on margins.des to generate design matrices for predicted probabilities. Instead, we constructed them manually. It is advisable to periodically search for new R packages that extend the current capabilities for these models.

11

Special Topics: Comparing between Models and Missing Data

In this chapter, we discuss a couple of additional important topics for regression models for limited, dependent variables. These topics did not have a natural placement in the body of the main description of these models, but we thought they were important enough topics to warrant a concluding chapter. Specifically, in this chapter, we consider comparing regression coefficients and model predictions from different model specifications, and how to address missing data among independent variables.

Comparing between Models

One should not directly compare the regression coefficients from different model specifications of a regression model for limited dependent variables. In the context of ordinary linear regression, adding additional variables to a model and assessing how the effect of a variable changes is relatively straightforward (Allison 1999). This is a common practice, particularly for evaluating mediation (Baron and Kenny 1986; Hayes 2017). The idea here is to show how adding an additional variable changes or alters the effect of an exogenous variable. Let us assume we are interested in estimating the effect of x on a binary y. The effect of x on y without control variables or mediators in the model is referred to as the *total effect* of x on y. After we control for additional variables, the effect of x on y is the *direct effect*. The difference between the total effect and the direct effect is the part of the total effect explained by adding the mediators to the model. This is generally termed the *indirect effect*.

In the context of regression models using a nonlinear link function (e.g., all of the models discussed in this book aside from OLS), the regression coefficients cannot be simply decomposed (MacKinnon and Dwyer 1993; Winship and Mare 1983). This is because the estimated regression coefficients and the model error variance are not separately identified. Instead, the estimated coefficients are a ratio of the true coefficients to a scale parameter that is defined by the error standard deviation (Amemiya 1981; Winship and Mare 1983). The scale parameter may vary from model to model, meaning that differences in coefficients, and therefore predicted values as well, may *not*

DOI: 10.1201/9781003029847-11

be directly compared between models. Both the coefficients and the scaling term varies from model to model, so differences in estimated coefficients or fitted values could be due to changes in coefficients or changes in scaling (Breen, Karlson, and Holm 2013, 2021).

We discuss two approaches for adjusting for differences in scaling across parameter estimates. First, we discuss the popular KHB method, developed by Karlson, Holm and Breen (2013, 2021). This method adjusts for scaling so that the decomposability of coefficients in the linear model is restored. This method is typically used to assess the change in coefficients as new variables are added to a model. The second approach we discuss focuses on the predicted or fitted values from different model specifications, as developed by Mize, Doan, and Scott Long (2019). This approach focuses on the average marginal effects and how they change as a result of different model specifications. This approach gets around issues of scaling by using seemingly unrelated estimation, which uses a sandwich/robust covariance matrix to combine parameter estimates (Weesie 2000). Because the parameter estimates are simultaneously generated from the same covariance matrix, the models share the same residual scaling. This makes the estimated coefficients and model fitted values comparable. We discuss both approaches in more detail below.

The KHB Method

The KHB method (Breen, Karlson, and Holm 2013, 2021) is rather straightforward to understand and implement. The reason the *total effect* does not decompose into *direct* and *indirect effects* in nonlinear regression models is because the inclusion of additional terms may alter the regression coefficients or the scaling of the residual. The KHB approach removes the influence of scaling by residualizing the additional independent variables from the original independent variables and including those residuals in the reduced model. Doing so means that the only difference between the full and reduced model is the systematic part of the additional variable that is associated with the outcome. The residual variation that affects the scaling of the residual is controlled in the reduced model.

Consider a simple three-variable system. Suppose we're interested in the effect of x on y and how z mediates that effect. The KHB approach entails regressing z on x, and including the model residual in the reduced model that estimates the total effect of x on y. This total effect, when controlling for z residualized from x, is decomposable just like in ordinary linear regression models.

We consider an example using the European Social Survey. To illustrate, we show output from the KHB function in R. To our knowledge, this function is not included in any package but is available online for users to copy and paste into their console.[1] We also include code to estimate these values without using the function (i.e., by manually computing the residual to be included in the reduced model). In our example, we are predicting whether the respondent thinks immigration is good for the economy. Our reduced model includes whether the respondent is racially minoritized, respondent religiousness, whether the respondent is a female, age, and employment status. The full model includes the variable "conservative" along with the variables in the reduced model. The print.khb function produces a nice summary of the KHB results:

```
> print.khb(k1)

KHB method
Model type: glm lm (logit)
Variables of interest:
minority, religious, female, age, self.emp, employee
Z variables (mediators): conservative

Summary of confounding
             Ratio Percentage Rescaling
minority  1.000181    0.018104    1.0093
religious 0.870549  -14.870077    1.0221
female    0.907798  -10.156655    1.0111
age       1.322009   24.357528    1.0099
self.emp  1.011188    1.106453    1.0128
employee  1.054074    5.130019    1.0127
----------------------------------------------------------------
minority :
          Estimate Std. Error z value Pr(>|z|)
Reduced 0.53169374 0.20010981  2.6570 0.007884 **
Full    0.53159748 0.20010960  2.6565 0.007895 **
Diff    0.00009626 0.02331014  0.0041 0.996705
----------------------------------------------------------------
religious :
          Estimate Std. Error z value Pr(>|z|)
Reduced  0.0484900  0.0165181  2.9356 0.003329 **
Full     0.0557005  0.0165994  3.3556 0.000792 ***
Diff    -0.0072105  0.0024508 -2.9421 0.003260 **
----------------------------------------------------------------
female :
          Estimate Std. Error z value Pr(>|z|)
Reduced -0.181330   0.095777 -1.8933  0.05832 .
Full    -0.199747   0.095868 -2.0836  0.03720 *
Diff     0.018417   0.012541  1.4685  0.14196
----------------------------------------------------------------
age :
           Estimate  Std. Error z value  Pr(>|z|)
Reduced -0.00988751  0.00272557 -3.6277  0.000286 ***
Full    -0.00747916  0.00275342 -2.7163  0.006601 **
Diff    -0.00240835  0.00055769 -4.3184 1.571e-05 ***
----------------------------------------------------------------
```

```
self.emp :
        Estimate Std. Error z value  Pr(>|z|)
Reduced 1.348188   0.321744  4.1902 2.787e-05 ***
Full    1.333271   0.321694  4.1445 3.405e-05 ***
Diff    0.014917   0.038960  0.3829   0.7018
---------------------------------------------------------------
employee :
        Estimate Std. Error z value  Pr(>|z|)
Reduced 1.358059   0.303235  4.4786 7.514e-06 ***
Full    1.288390   0.303197  4.2493 2.144e-05 ***
Diff    0.069669   0.038560  1.8068   0.0708 .
```

After describing the call, `print.khb` first provides a summary how much the coefficients change with and without including the residuals (Breen, Karlson, and Holm 2013, 2021), labeled "Summary of confounding." After the summary of confounding is a model summary for each variable. Estimates include the estimated regression coefficient for the reduced model (that includes the residual), the full model, and the difference between the estimated coefficients. Results show that conservative mediates a significant share of the effect age has on whether the respondent thinks immigration is good for the economy. Specifically, the reduced model shows that as respondents get older, they are less likely to think immigration is good for the economy. After we control for political conservatism, the effect of age is significantly reduced (difference = −.002, s.e. = .0006, $p < .001$). A significant reduction in a coefficient, as in the previous sentence, is formal evidence of mediation in the context of a regression model for a limited dependent variable.

More generally, the KHB method restores the decomposability of ordinary linear regression coefficients to the entire family of the GLM (Breen, Karlson, and Holm 2013, 2021). Such adjustments are required to compare coefficients between different model specifications when the outcome is not continuous. In the next section, we describe another logically similar approach to comparing between regression models for limited dependent variables.

Comparing Marginal Effects

Mize, Doan, and Long (2019) argue for making model comparisons on the response metric, illustrating how fitted values change as the model specification changes. This is consistent with our emphasis on illustrating the implications of regression models for limited dependent variables using fitted values or the response scale itself, rather than mathematically manipulated regression coefficients.

By moving to the response metric, Mize, Doan, and Long (2019) get around the issue of residual scaling because the assumed error terms do not affect the predicted probabilities (Long 1997). They use seemingly unrelated estimation (suest), which exploits a nonstandard way of using the sandwich estimator to combine separate results into a single covariance matrix to allow for tests of significance. Once the models are estimated and combined, Mize, Doan, and Long advocate for computing marginal effects for each regression equation, and then comparing those marginal effects to illustrate how model specification alters the marginal effects. For example, controlling for a mediator should reduce the AME of variables the mediator is thought to explain.

Practically, Mize, Doan, and Long (2019) implement their method in the context of a generalized structural equation model (SEM) in Stata. The SEM uses a single covariance matrix to generate the parameter estimates. Stata then draws margins from each equation and tests the relationship between margins.

We tried three approaches to implement this method in R. First, we used SEM, as the authors do in Stata. Unfortunately, drawing margins from SEM objects in the lavaan package (Rosseel 2012) is not as straightforward as we would like. Second, we tried to fit a seemingly unrelated regression using nlsystemfit. We found this to be unstable and quite inefficient (i.e., run time was very long). Third, rather than fit the SUR as a system of simultaneously estimated regression equations, we use KHB principles to generate the marginal effects that we can then compare using simulation-based inference. As Williams and Jorgensen (2022: 10) note, "the Mize, Doan, and Long approach could…be tweaked to use the KHB Reduced Model," and "our experience suggests that differences between the two approaches tend to be slight." We concur and note that this is the approach we advocate for in terms of estimation in R at this point.

Consider our example above. If we want to know how conservative alters the marginal effect of age on attitudes toward immigration, we regress conservative on age and retain the model residual. Next, we regress attitudes toward immigration on age controlling for the residual of conservative on age. We compute the marginal effect of age on attitudes toward immigration, net of the residual of conservative on age. Next, we estimate another logistic regression, regressing attitudes on age and conservative, and compute the marginal effect of age in this model. Now we have two marginal effects: one estimate without controlling for conservative (although we have controlled for residual scaling due to conservative) and the other estimate after controlling for conservative. We then compare the AMEs using the compare. margins function. Importantly, generating the marginal effects in this way means that we cannot use the Delta method to compute the standard error of the change in marginal effects. This is because the marginal effects come from separately estimated models. The compare.margins function relies on simulations and only assumes that each marginal effect is normally distributed.

We provide two examples in the online supplement. In the first, we compute AMEs for each predictor across models, and test for differences between them. In the second, we compute MEMs for age. Figure 11.1 shows the MEM for age in two different logistic regression models. In the first, we control for conservative residualized from the other variables in the model. In the

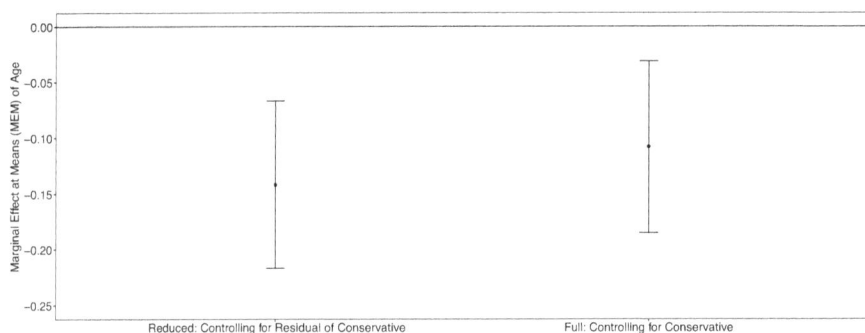

FIGURE 11.1
Marginal effects at means for age.

second, we control for conservative. The figure therefore illustrates the change in the MEM of age after controlling for conservative, which is .034 ($p=.267$).

Missing Data

Another common problem faced by analysts is missing data. By default, statistical platforms use listwise deletion to address item nonresponse (i.e., missing values on some variables for some cases). Listwise deletion results in inefficient parameter estimates (Allison 2001). If you can assume that missing values are missing at random (Allison 2001), a common approach to addressing item nonresponse is multiple imputation (Little and Rubin 2019; Rubin 2004), particularly in the context of the models discussed in this book.[2]

If you have reasons to believe that item nonresponse is not a random process or if you do not have a sense of how item nonresponse occurred, you should examine your missing data patterns. A first step in this direction is to create a dummy variable to indicate whether cases are missing on a particular variable. Then, other variables can be used to predict the missing variable for cases that are or are not missing. Suppose, for example, that lower income respondents are less likely to report their income. Any variable that is positively correlated with income would show the following pattern: the conditional mean for respondents with observed income values will be higher than the conditional mean for respondents with unobserved income values. This pattern is due to the correlation: those with lower incomes, and therefore an increased likelihood of missing values, will have lower values on the second variable. If item nonresponse is truly a random process, there will not be differences in the conditional means/proportions of independent variables by the missingness status of other variables. We also refer readers to the `finalfit` package (Harrison, Drake, and Ots 2022), and in particular, its `missing _ plot` and `missing _ pattern` functions. These are useful for in-depth exploration of missing data patterns in higher dimensions.

If we can assume that item nonresponse is a random process, multiple imputation is the best strategy for dealing with item nonresponse of independent variables for the models discussed in this book. Multiple imputation entails (*i*) imputing missing values with some random error, (*ii*) repeating this process many (usually 20) times, (*iii*) fitting our desired model or models to the multiple data sets, and then (*iv*) aggregating over the model estimates to define a single set of parameter estimates (e.g., coefficients and their standard errors). In terms of how to impute the missing values, multiple imputation for chained equations (MICE) is the currently preferred strategy (White, Royston, and Wood 2011). Imputed values are generated from a series of (chained) regression models, where the imputation of each variable can be a function of whatever predictors you want to include, and values are imputed on the response scale of each predictor. You specify the link function for each predictor variable, so binary variables can be imputed using logistic regression, continuous variables can be imputed using linear regression, multinomial variables can be imputed using multinomial logistic regression, and so on.

R's MICE package (van Buuren and Groothuis-Oudshoorn 2011) implements MICE. We refer readers to the supporting material for the package for more details, but here we point out how to implement steps $i - iv$ from above. We do so in the context of our first logistic regression model from Chapter 5. Before describing the details of how to implement the MICE package for multiple imputation, we describe missing data patterns. Table 11.1 shows the means/proportions, standard deviations, effective sample sizes, and percent missing for each variable. There is .5% of cases that are missing on the dependent variable. We do not consider models for sample selection bias on the dependent variable, but instead refer readers to the `sampleSelection` package (Toomet and Henningsen 2008). As shown in Table 11.1, there are missing values on religiousness for .2% of respondents, on minority for .5% of respondents, and on age for .7% of respondents. While the amount of

TABLE 11.1

Descriptive Statistics for the Variables in the Logistic Regression Model

Variable	Mean	SD	N	% Missing
Feel Safe at Night	0.754		2193	0.005
Religiousness	3.606	3.074	2200	0.002
Racially Minoritized	0.078		2194	0.005
Female	0.547		2204	0.000
Age	53.274	18.406	2188	0.007
Employee	0.797		2204	0.000
Self-Employed	0.169		2204	0.000
Unemployed	0.034		2204	0.000

missing data in this example is quite small, the principles generalize regardless of the amount.

To begin, we specify a MICE statement asking for it to impute missing values zero times. This returns default settings and provides objects to manipulate if we want to alter the default settings. There are two important settings that we may want to alter. The first is the `method`, which defines the link function used for each predictor variable when imputing missing values. We prefer to use `norm` for continuous variables, `logreg` for binary variables (logistic regression), and `polyreg` for categorical variables with more than two states (multinomial logistic regression). MICE supports many other functions as well (van Buuren and Groothuis-Oudshoorn 2011). The second important setting is the `predictorMatrix`. This defines which variables are used to impute missing values on each predictor. Specifically, you can save the `predictorMatrix`, alter it, and give it to a subsequent MICE function to take control of which variables are used to predict missing values on each predictor. By default, MICE uses every variable to predict missing values on every other variable. Changing values in the rows of `predictorMatrix` from 1 to 0 means to exclude the variable in that column from predicting missing values on the variable in that row. We provide an example of this in the supporting material.

After you define a `mids` object, which is the result of the MICE function, for a full set of multiple imputations, the next step is to estimate your regression model on the imputed data sets. We use the `with` function to do so. In this context, this function takes the `mids` object and the model statement, and estimates the model on each of the imputed data files. Finally, the `pool` function aggregates the model estimates into a single set of pooled estimates. The regression coefficients are the mean of the regression coefficients across imputed data sets. The standard errors are aggregated according to Rubin's rule (Rubin 2004). The code and output below show the implementation of a default `predictorMatrix`, but a custom `method` statement for the logistic regression model is of interest. First is the MICE statement. Then, we fit the same logistic regression model to each imputed data set using the with function. The pool function aggregates the results for us, and then we ask for a summary of the pooled estimates. These estimates are more efficient than the estimates in Chapter 5 that rely on listwise deletion.

```
> out<-mice(X2,m=20,seed=1982,method=c("","norm","logreg","logreg",
"norm","polyreg"),printFlag = FALSE)

> fits<-with(out,glm(safe ~ religious + minority  + female + age + self.
emp + employee,data=X,family=binomial))

> est<-pool(fits)

> summary(est)
```

```
          term       estimate    std.error  statistic        df      p.value
1 (Intercept)    1.604911098  0.294422712   5.4510438  2153.785  5.580332e-08
2    religious  -0.018508071  0.017781583  -1.0408562  2153.785  2.980591e-01
3     minority  -0.168608905  0.194664693  -0.8661504  2153.785  3.865042e-01
4       female  -1.033937698  0.112276590  -9.2088449  2153.785  0.000000e+00
5          age  -0.007040938  0.003003321  -2.3443841  2153.785  1.914895e-02
6     self.emp   1.016186376  0.297467128   3.4161300  2153.785  6.469654e-04
7     employee   0.573451413  0.263415344   2.1769856  2153.785  2.958990e-02
```

In the example above, we applied MICE to a logistic regression model. One of the most appealing features of MICE is that it can be used regardless of the desired model. MICE can be used to adjust for item nonresponse for all the models discussed in this book, and for almost all models in general. To do so, we simply change the model statement in the `with` function.

Margins with MICE

As we have noted throughout this book, illustrating the implications of a regression model for limited dependent variables is often done on the response metric. For example, to illustrate the effect of age on whether respondents think immigration is good for the economy, we would want to show how the probability of the outcome varies with age. When using MICE, we have m possible fitted values, where m refers to the number of imputed data sets. The solution in this case is to estimate the margin or fitted value for each of the imputed data sets, and then aggregate over them using the same procedures we use to aggregate the model coefficients. The fitted value is the mean across the set of m fitted values, and the standard error for the fitted value can be obtained by applying Rubin's rule to the standard errors for each margin. Generally, Rubin's rule is applicable to variables that are assumed to follow a normal distribution, including model predictions and marginal effects (Allison 2009).

Using our example, we computed the average marginal effect (AME) for female. To do so, we first compute the AME of female for the first fully-imputed data set. The code for this is as follows:

```
margs<-data.frame(summary(margins(fits$analyses[[1]],variables
=c("female"))))
```

Note that we are indexing into `fits` to draw margins from a model labeled `analysis`. We then create a loop, which increases i from 2 to 20 each time it passes. For each pass, we compute the AME for the ith imputed data set (`margi`) and append it to the existing estimates (`margs`). At the end of the loop, we have an object with 20 sets of AMEs and their standard errors. We aggregate these values using `tidyverse`'s `summarize` function, applying the `mean` and `rubins.rule` functions, respectively. Here is the loop and aggregation:

```
> for(i in 2:20){
+    margsi<-data.frame(summary(margins(fits$analyses[[i]],vari
ables=c("female"))))
+    margs<-rbind(margs,margsi)}

> margs %>% summarize(AME=mean(AME),SE=rubins.rule(SE))

          AME         SE
1 -0.1801204 0.01853468
```

The *p*-value for the AME can be obtained with the dnorm function, that is,

```
> dnorm(-.1801204/.01853468)
[1] 1.240342e-21
```

In this case, the margins say that females are .18 less likely to report that immigration is good for the economy than males, and this effect is significant ($p < .001$). More generally, missing data due to item nonresponse may make your parameter estimates inefficient. This is true of both the estimated regression coefficients and any post-estimation results. As such, we suggest using multiple imputation, along with the principles described earlier in this book, to develop your regression models for limited dependent variables when you have item nonresponse among your predictor variables.

Conclusion

In this chapter, we discussed two important extensions to regression models for limited dependent variables, and the issues discussed here are relevant to all the models in this book. Comparing across models requires great care, because differences in parameter estimates between different model specifications may be driven by not only true differences in the underlying coefficients but also differences in residual scaling from model to model (Amemiya 1981; Winship and Mare 1983). We described two approaches to adjust for residual scaling: the KHB method and using suest to compare margins. While suest is comparatively underdeveloped in R, the generality of the KHB method enables us to generate comparable margins. It takes a little bit of programming to do so (i.e., generating the residual to include in the reduced model) but is ultimately estimable. We hope that other scholars will fill in some of these gaps in the coming years so that the full suite of modern methods for alternative-specific outcomes (Chapter 10) and comparing between models is readily available to all, for free, in R.

The second issue we addressed in this chapter is item nonresponse. We provided a simple heuristic for assessing whether item nonresponse is nonrandom and noted additional resources in this area. If item nonresponse can be assumed to be a random process, we then illustrated how to implement multiple imputation for chained equations. The MICE package makes this straightforward to implement, but margins from multiply imputed data need to be aggregated. We showed how to use a loop to estimate the margins and then use simple functions to aggregate the results.

Notes

1. The functions are currently (November 2, 2022) available here: https://rdrr.io/rforge/khb/man/khb.html
2. In the case of OLS regression, Full Information Maximum Likelihood (FIML) estimation is preferred (e.g., Allison 2009b). Unfortunately, this estimator is not widely available for regression models of limited dependent variables .

References

Agresti, Alan. 2003. *Categorical Data Analysis*. New York: John Wiley & Sons.

Ai, Chunrong, and Edward C. Norton. 2003. "Interaction Terms in Logit and Probit Models." *Economics Letters* 80(1):123–129.

Akaike, Hirotugu. 1974. "A New Look at the Statistical Model Identification." *IEEE Transactions on Automatic Control* 19(6):716–723.

Allison, Paul D. 1999. *Multiple Regression: A Primer*. Thousand Oaks, CA: Pine Forge Press.

Allison, Paul D. 2001. *Missing Data*. Thousand Oaks, CA: Sage.

Allison, Paul D. 2009a. *Fixed Effects Regression Models*. Thousand Oaks, CA: Sage.

Allison, Paul D. 2009b. "Missing Data." Pp. 72–89 in *The Sage Handbook of Quantitative Methods in Psychology*. Thousand Oaks, CA: Sage.

Allison, Paul D., and Nicholas A. Christakis. 1994. "Logit Models for Sets of Ranked Items." *Sociological Methodology* 24: 199–228.

Amemiya, Takeshi. 1981. "Qualitative Response Models: A Survey." *Journal of Economic Literature* 19(4):1483–1536.

Arel-Bundock, Vincent. 2022. *Marginaleffects: Marginal Effects, Marginal Means, Predictions, and Contrasts*. CRAN.

Baron, Reuben M., and David A. Kenny. 1986. "The Moderator–Mediator Variable Distinction in Social Psychological Research: Conceptual, Strategic, and Statistical Considerations." *Journal of Personality and Social Psychology* 51(6):1173.

Beller, Emily. 2009. "Bringing Intergenerational Social Mobility Research into the Twenty-First Century: Why Mothers Matter." *American Sociological Review* 74(4):507–528.

Belsley, David A., Edwin Kuh, and Roy E. Welsch. 2005. *Regression Diagnostics: Identifying Influential Data and Sources of Collinearity*. New York: John Wiley & Sons.

Brant, Rollin. 1990. "Assessing Proportionality in the Proportional Odds Model for Ordinal Logistic Regression." *Biometrics* 46(4): 1171–1178.

Breen, Richard, Kristian Bernt Karlson, and Anders Holm. 2013. "Total, Direct, and Indirect Effects in Logit and Probit Models." *Sociological Methods & Research* 42(2): 164–191.

Breen, Richard, Kristian Bernt Karlson, and Anders Holm. 2021. "A Note on a Reformulation of the KHB Method." *Sociological Methods & Research* 50(2):901–912.

Breiger, Ronald L. 1981. "The Social Class Structure of Occupational Mobility." *American Journal of Sociology* 87(3):578–611.

Buehler, Ralph, and Andrea Hamre. 2015. "The Multimodal Majority? Driving, Walking, Cycling, and Public Transportation Use among American Adults." *Transportation* 42(6):1081–1101.

van Buuren, Stef, and Karin Groothuis-Oudshoorn. 2011. "Mice: Multivariate Imputation by Chained Equations in R." *Journal of Statistical Software* 45(3):1–67. doi: 10.18637/jss.v045.i03.

Cameron, A. Colin, and Pravin K. Trivedi. 2013. *Regression Analysis of Count Data*. Vol. 53. New York: Cambridge University Press.

Carstensen, Bendix, Martyn Plummer, Esa Laara, and Michael Hills. 2015. "Epi: A Package for Statistical Analysis in Epidemiology. R Package Version 1.1. 67." CRAN.

Chen, Guo, and Hiroki Tsurumi. 2010. "Probit and Logit Model Selection." *Communications in Statistics—Theory and Methods* 40(1):159–75.

Cheng, Simon, and J. Scott Long. 2007. "Testing for IIA in the Multinomial Logit Model." *Sociological Methods & Research* 35(4):583–600.

Clogg, Clifford C., and Scott R. Eliason. 1987. "Some Common Problems in Log-Linear Analysis." *Sociological Methods & Research* 16(1):8–44.

Clogg, Clifford C., and Edward S. Shihadeh. 1994. *Statistical Models for Ordinal Variables*. Vol. 4. Thousand Oaks, CA: Sage.

Collett, David. 2002. *Modelling Binary Data*. New York: Chapman & Hall/CRC Press.

Coxe, Stefany, Stephen G. West, and Leona S. Aiken. 2009. "The Analysis of Count Data: A Gentle Introduction to Poisson Regression and Its Alternatives." *Journal of Personality Assessment* 91(2):121–136.

Crawley, Michael J. 2012. *The R Book*. New York: John Wiley & Sons.

Croissant, Yves. 2012. "Estimation of Multinomial Logit Models in R: The Mlogit Packages." *R Package Version 0.2-2*. URL: Http://Cran. r-Project. Org/Web/ Packages/Mlogit/Vignettes/Mlogit. Pdf. CRAN.

Croissant, Yves. 2020. "Estimation of Random Utility Models in R: The Mlogit Package." *Journal of Statistical Software* 95(11):1–41. doi: 10.18637/jss.v095.i11.

Dow, Jay K., and James W. Endersby. 2004. "Multinomial Probit and Multinomial Logit: A Comparison of Choice Models for Voting Research." *Electoral Studies* 23(1):107–122.

Efron, Bradley, and Robert J. Tibshirani. 1994. *An Introduction to the Bootstrap*. New York: Chapman & Hall/CRC press.

Eliason, Scott R. 1993. *Maximum Likelihood Estimation: Logic and Practice*. Thousand Oaks, CA: Sage.

Feng, Cindy Xin. 2021. "A Comparison of Zero-Inflated and Hurdle Models for Modeling Zero-Inflated Count Data." *Journal of Statistical Distributions and Applications* 8(1): 1–19.

Fisher, Ronald A. 1922. "On the Interpretation of $\chi 2$ from Contingency Tables, and the Calculation of P." *Journal of the Royal Statistical Society* 85(1):87–94.

Forbes, Catherine, Merran Evans, Nicholas Hastings, and Brian Peacock. 2011. *Statistical Distributions*. Vol. 4. New York: Wiley New York.

Fox, John. 1997. *Applied Regression Analysis, Linear Models, and Related Methods*. Thousand Oaks, CA: Sage.

Fox, John, and Sanford Weisberg. 2019. *An R Companion to Applied Regression*. Third. Thousand Oaks, CA: Sage.

Fry, Tim RL, and Mark N. Harris. 1996. "A Monte Carlo Study of Tests for the Independence of Irrelevant Alternatives Property." *Transportation Research Part B: Methodological* 30(1):19–30.

Fry, Tim RL, and Mark N. Harris. 1998. "Testing for Independence of Irrelevant Alternatives: Some Empirical Results." *Sociological Methods & Research* 26(3):401–423.

Goodman, Leo A. 1968. "The Analysis of Cross-Classified Data: Independence, Quasi-Independence, and Interactions in Contingency Tables with or without Missing Entries: Ra Fisher Memorial Lecture." *Journal of the American Statistical Association* 63(324):1091–1131.

Goodman, Leo A. 1979. "Simple Models for the Analysis of Association in Cross-Classifications Having Ordered Categories." *Journal of the American Statistical Association* 74(367):537–552.

Goodman, Leo A. 1981a. "Association Models and Canonical Correlation in the Analysis of Cross-Classifications Having Ordered Categories." *Journal of the American Statistical Association* 76(374):320–334.

Goodman, Leo A. 1981b. "Association Models and the Bivariate Normal for Contingency Tables with Ordered Categories." *Biometrika* 68(2):347–355.

Greene, David L., and Donald W. Jones. 1997. *The Full Costs and Benefits of Transportation: Contributions to Theory, Method and Measurement; with 62 Tables*. New York: Springer Science & Business Media.

Greene, William H. 2003. *Econometric Analysis*. Columbus, OH: Pearson Education.

Greil, Arthur L., Kathleen Slauson-Blevins, and Julia McQuillan. 2010. "The Experience of Infertility: A Review of Recent Literature." *Sociology of Health & Illness* 32(1):140–162.

Gunst, Richard F., and Robert L. Mason. 2018. *Regression Analysis and Its Application: A Data-Oriented Approach*. New York: Chapman & Hall/CRC Press.

Haberman, Shelby J. 1981. "Tests for Independence in Two-Way Contingency Tables Based on Canonical Correlation and on Linear-by-Linear Interaction." *The Annals of Statistics* 3(2): 1178–1186.

Hardin, James W. and Joseph Hilbe. 2007. *Generalized Linear Models and Extensions*. Austin, TX: Stata Press.

Harrison, Ewen, Tom Drake, and Riinu Ots. 2022. *Finalfit: Quickly Create Elegant Regression Results Tables and Plots When Modelling*. CRAN.

Hausman, Jerry, and Daniel McFadden. 1984. "Specification Tests for the Multinomial Logit Model." *Econometrica: Journal of the Econometric Society* 52(5): 1219–1240.

Hayes, Andrew F. 2017. *Introduction to Mediation, Moderation, and Conditional Process Analysis: A Regression-Based Approach*. New York: Guilford Publications.

Healy, Kieran. 2018. *Data Visualization: A Practical Introduction*. Princeton, NJ: Princeton University Press.

Hilbe, Joseph M. 2011. *Negative Binomial Regression*. New York: Cambridge University Press.

Hlavac, Marek. 2015. "Stargazer: Beautiful LATEX, HTML and ASCII Tables from R Statistical Output." CRAN.

Hosmer, David W., and Stanley Lemeshow. 2000. *Applied Logistic Regression*. New York: John Wiley & Sons.

Jackman, Simon, Alex Tahk, Achim Zeileis, Christina Maimone, Jim Fearon, Zoe Meers, Maintainer Simon Jackman, and MASS Imports. 2015. "Package 'Pscl.'" *Political Science Computational Laboratory* 18(04.2017). CRAN.

Kassambara, Alboukadel. 2020. *Ggpubr: "ggplot2" Based Publication Ready Plots*. CRAN.

King, Gary, Michael Tomz, and Jason Wittenberg. 2000. "Making the Most of Statistical Analyses: Improving Interpretation and Presentation." *American Journal of Political Science* 44(2): 347–361.

Knoke, David, and Peter J. Burke. 1980. *Log-Linear Models*. Thousand Oaks, CA: Sage.

Kuha, Jouni. 2004. "AIC and BIC: Comparisons of Assumptions and Performance." *Sociological Methods & Research* 33(2):188–229.

Lambert, Diane. 1992. "Zero-Inflated Poisson Regression, with an Application to Defects in Manufacturing." *Technometrics* 34(1):1–14.

Leeper, Thomas J. 2017. "Interpreting Regression Results Using Average Marginal Effects with R's Margins." *Comprehensive R Archive Network (CRAN)*, 1–32.

Leeper, Thomas J. 2021. *Margins: Marginal Effects for Model Objects*. CRAN.

Lenth, Russell, Henrik Singmann, Jonathon Love, Paul Buerkner, and Maxime Herve. 2021. "Emmeans: Estimated Marginal Means, Aka Least-Squares Means." *R Package Version* 1(1):3.

Lesnoff, M., Lancelot, R. 2012. *Aod: Analysis of Overdispersed Data*. CRAN.

Little, Roderick JA, and Donald B. Rubin. 2019. *Statistical Analysis with Missing Data*. Vol. 793. New York: John Wiley & Sons.

Logan, John A. 1983. "A Multivariate Model for Mobility Tables." *American Journal of Sociology* 89(2):324–349.

Long, J. S. 1997. *Regression Models for Categorical and Limited Dependent Variables*. Thousand Oaks, CA: Sage.

Long, J. Scott, and Jeremy Freese. 2006. *Regression Models for Categorical Dependent Variables Using Stata*. Vol. 7. Austin, TX: Stata Press.

Long, J. Scott, and Sarah A. Mustillo. 2021. "Using Predictions and Marginal Effects to Compare Groups in Regression Models for Binary Outcomes." *Sociological Methods & Research* 50(3): 1284–1320.

Lüdecke, Daniel. 2018. "Ggeffects: Tidy Data Frames of Marginal Effects from Regression Models." *Journal of Open Source Software* 3(26):772.

MacKinnon, David P., and James H. Dwyer. 1993. "Estimating Mediated Effects in Prevention Studies." *Evaluation Review* 17(2): 144–158.

MacKinnon, David P., Jennifer L. Krull, and Chondra M. Lockwood. 2000. "Equivalence of the Mediation, Confounding and Suppression Effect." *Prevention Science* 1(4): 173–181.

McFadden, Daniel. 1973. "Conditional Logit Analysis of Qualitative Choice Behavior." Pp. 105–142 in *Frontiers in Econometrics*. New York: Academic Press.

Melamed, David. 2015. "Communities of Classes: A Network Approach to Social Mobility." *Research in Social Stratification and Mobility* 41:56–65.

Mize, Trenton D. 2019. "Best Practices for Estimating, Interpreting, and Presenting Nonlinear Interaction Effects." *Sociological Science* 6(4):81–117.

Mize, Trenton D., Long Doan, and J. Scott Long. 2019. "A General Framework for Comparing Predictions and Marginal Effects across Models." *Sociological Methodology* 49(1):152–189.

Moore, Julia. 2017. "Facets of Agency in Stories of Transforming from Childless by Choice to Mother." *Journal of Marriage and Family* 79(4):1144–1159.

Mullahy, John. 1986. "Specification and Testing of Some Modified Count Data Models." *Journal of Econometrics* 33(3):341–365.

Neter, John, Michael H. Kutner, Christopher J. Nachtsheim, and William Wasserman. 1996. *Applied Linear Statistical Models*. Ann Arbor, MI: Irwin Press.

Paap, Richard, and Philip Hans Franses. 2000. "A Dynamic Multinomial Probit Model for Brand Choice with Different Long-Run and Short-Run Effects of Marketing-Mix Variables." *Journal of Applied Econometrics* 15(6): 717–744.

Pearson, Karl. 1900. "X. On the Criterion That a given System of Deviations from the Probable in the Case of a Correlated System of Variables Is Such That It Can Be Reasonably Supposed to Have Arisen from Random Sampling." *The London, Edinburgh, and Dublin Philosophical Magazine and Journal of Science* 50(302):157–175.

Peterson, Bercedis, and Frank E. Harrell Jr. 1990. "Partial Proportional Odds Models for Ordinal Response Variables." *Journal of the Royal Statistical Society: Series C (Applied Statistics)* 39(2):205–217.

Powers, Daniel, and Yu Xie. 2008. *Statistical Methods for Categorical Data Analysis*. New York: Emerald Group Publishing.

Pregibon, Daryl. 1981. "Logistic Regression Diagnostics." *The Annals of Statistics* 9(4):705–724.

Raftery, Adrian E. 1995. "Bayesian Model Selection in Social Research." *Sociological Methodology* 25(1995): 111–163.

Ray, Paramesh. 1973. "Independence of Irrelevant Alternatives." *Econometrica: Journal of the Econometric Society* 41(4): 987–991.

Ripley, Brian, Bill Venables, Douglas M. Bates, Kurt Hornik, Albrecht Gebhardt, David Firth, and Maintainer Brian Ripley. 2013. "Package 'Mass.'" 538:113–120 CRAN.

Ripley, Brian, William Venables, and Maintainer Brian Ripley. 2016. "Package 'Nnet.'" *R Package Version* 7(3–12):700. CRAN.

Robinson, David, Alex Hayes, and Simon Couch. 2022. *Broom: Convert Statistical Objects into Tidy Tibbles*. CRAN.

Rosseel, Yves. 2012. "Lavaan: An R Package for Structural Equation Modeling." *Journal of Statistical Software* 48(2):1–36. doi: 10.18637/jss.v048.i02.

Rubin, Donald B. 2004. *Multiple Imputation for Nonresponse in Surveys*. Vol. 81. New York: John Wiley & Sons.

Schlegel, Benjamin, and Marco Steenbergen. 2020. *Brant: Test for Parallel Regression Assumption*. CRAN.

Schwarz, Gideon. 1978. "Estimating the Dimension of a Model." *The Annals of Statistics* 6(2): 461–464.

Searle, Shayle R., F. Michael Speed, and George A. Milliken. 1980. "Population Marginal Means in the Linear Model: An Alternative to Least Squares Means." *The American Statistician* 34(4):216–221.

Snijders, Tom AB, and Roel J. Bosker. 2011. *Multilevel Analysis: An Introduction to Basic and Advanced Multilevel Modeling*. Thousand Oaks, CA: Sage.

Sobel, Michael E., Michael Hout, and Otis Dudley Duncan. 1985. "Exchange, Structure, and Symmetry in Occupational Mobility." *American Journal of Sociology* 91(2): 359–372.

Team, R. Core, Roger Bivand, Vincent J. Carey, Saikat DebRoy, Stephen Eglen, Rajarshi Guha, Swetlana Herbrandt, Nicholas Lewin-Koh, Mark Myatt, and Michael Nelson. 2022. "Package 'Foreign.'" CRAN.

Therneau, Terry M. 2022. *A Package for Survival Analysis in R*. CRAN.

Timoneda, Joan C. 2021. "Estimating Group Fixed Effects in Panel Data with a Binary Dependent Variable: How the LPM Outperforms Logistic Regression in Rare Events Data." *Social Science Research* 93: 102486.

Toomet, Ott, and Arne Henningsen. 2008. "Sample Selection Models in R: Package SampleSelection." *Journal of Statistical Software* 27(7): 1–23.

Train, Kenneth E. 2009. *Discrete Choice Methods with Simulation*. New York: Cambridge University Press.

Verzani, John. 2011. *Getting Started with RStudio*. New York: O'Reilly Media, Inc.

Vuong, Quang H. 1989. "Likelihood Ratio Tests for Model Selection and Non-Nested Hypotheses." *Econometrica: Journal of the Econometric Society* 57(2): 307–333.

Weesie, Jeroen. 2000. "Seemlingly Unrelated Estimation and the Cluster-Adjusted Sandwich Estimator." *Stata Technical Bulletin* 9(52).

White, Ian R., Patrick Royston, and Angela M. Wood. 2011. "Multiple Imputation Using Chained Equations: Issues and Guidance for Practice." *Statistics in Medicine* 30(4): 377–399.

Wickham, Hadley, Mara Averick, Jennifer Bryan, Winston Chang, Lucy D'Agostino McGowan, Romain François, Garrett Grolemund, Alex Hayes, Lionel Henry, and Jim Hester. 2019. "Welcome to the Tidyverse." *Journal of Open Source Software* 4(43): 1686.

Wickham, Hadley, Winston Chang, and Maintainer Hadley Wickham. 2016. "Package 'Ggplot2.'" *Create Elegant Data Visualisations Using the Grammar of Graphics. Version* 2(1): 1–189.

Wickham, Hadley, and Garrett Grolemund. 2016. *R for Data Science: Import, Tidy, Transform, Visualize, and Model Data*. New York: O'Reilly Media, Inc.

Wickham, Hadley, Jim Hester, Winston Chang, and Jennifer Bryan. 2021. *Devtools: Tools to Make Developing R Packages Easier*. CRAN.

Williams, Richard. 2005. "Gologit2: A Program for Generalized Logistic Regression/ Partial Proportional Odds Models for Ordinal Dependent Variables." *Stata, Gologit2 Manual*.

Williams, Richard, and Abigail Jorgensen. 2022. "Comparing Logit & Probit Coefficients between Nested Models." *Social Science Research* 109(4): 102802.

Winship, Christopher, and Robert D. Mare. 1983. "Structural Equations and Path Analysis for Discrete Data." *American Journal of Sociology* 89(1):54–110.

Wooldridge, Jeffrey M. 2010. *Econometric Analysis of Cross Section and Panel Data*. Boston, MA: MIT press.

Yates, Frank. 1934. "Contingency Tables Involving Small Numbers and the $\chi 2$ Test." *Supplement to the Journal of the Royal Statistical Society* 1(2):217–235.

Yee, Thomas W., Jakub Stoklosa, and Richard M. Huggins. 2015. "The VGAM Package for Capture-Recapture Data Using the Conditional Likelihood." *Journal of Statistical Software* 65(5): 1–33.

Zeileis, Achim, Christian Kleiber, and Simon Jackman. 2008. "Regression Models for Count Data in R." *Journal of Statistical Software* 27(8):1–25.

Index

For Product Safety Concerns and Information please contact our EU
representative GPSR@taylorandfrancis.com
Taylor & Francis Verlag GmbH, Kaufingerstraße 24, 80331 München, Germany

www.ingramcontent.com/pod-product-compliance
Lightning Source LLC
Chambersburg PA
CBHW060405220326
41598CB00023B/3028

9 7 8 1 0 3 2 5 0 9 5 1 8